Every time I open my eyes, I see things

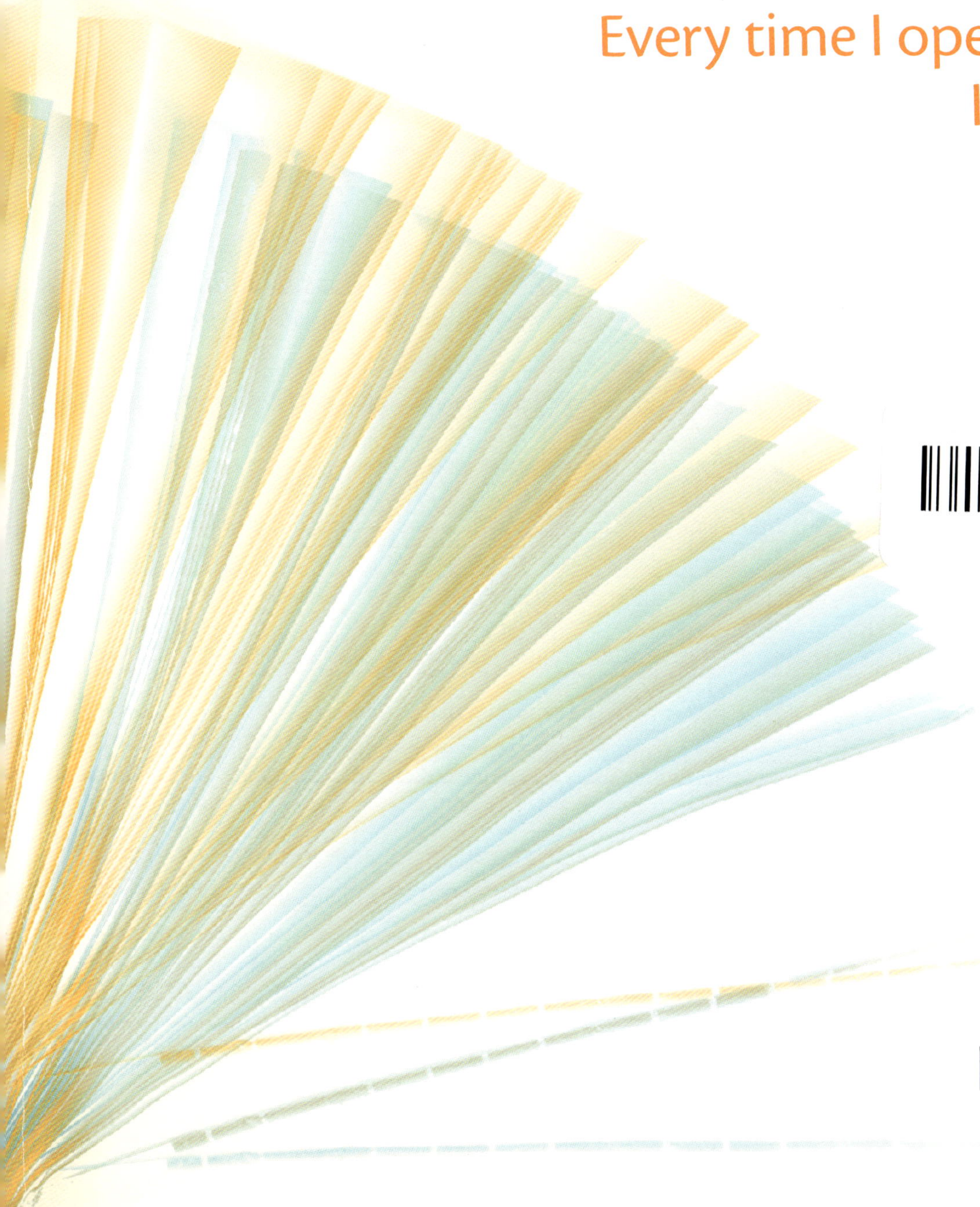

Doris Vila

A survey

The art of Doris Vila does what great art is supposed to do. Her work is technically wonderful. Employing new and traditional technologies, she creates extraordinary images with a strong conceptual element. Her marrying of traditional video with holography is especially significant and appealing.

These pieces on one level razzle-dazzle the viewer with dramatic three-dimensional effects and motion. While they might tease the eyes, their effectiveness lies in the way in which they tease the mind. One looks at and into the hologram and is entertained. Yet one finds oneself trying to analyze the narrative, searching for meanings and asking questions.

Art that endures is always the art that is laden with ideas. Doris Vila's art appears to offer more questions than answers. She has a way of inviting the viewers in for a look and then lures them into a mysterious world of the improbable. She has a mastery over her medium and understands in the deepest sense why art exists and why, in its purest state, it can move us to higher levels of understanding.

Louis A. Zona
Director
Butler Institute of American Art

I imagine tearing up a book, throwing it up in the air and trying to read the scraps of pages as they fall to the ground. This book is what you might see if the pieces of work landed just right. Doris Vila, Brooklyn, 2008

Vilagrams

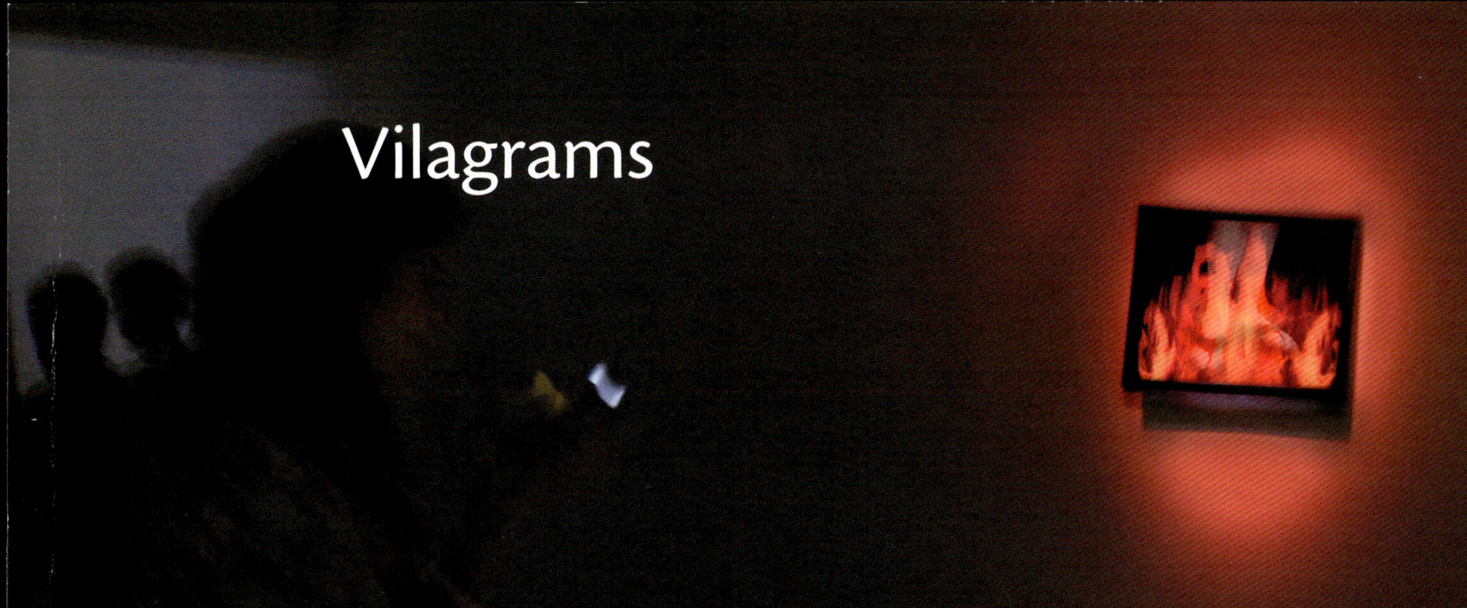

Holography and video both arrive in the late 1960s, each enabling visual artists to realize images that could not be done before. While millions around the world have made videos, nearly all of them amateur, less than one hundred holographers are capable of realizing holograms wholly by themselves.

Holography is very hard to do. I know because I couldn't do it by myself and so needed help first from Hart Perry and then from Fred Unterseher, two of the pros, to realize my own holograms. If videographers nowadays can see their final images on a screen before them, the holographer must negotiate several daunting obstacles before seeing anything. Typically, in classic split-beam holography, if the holographer screwed up, the result was not a blurred image, as in photography, say, but nothing, absolutely nothing.

Soon after I saw my first videotape and first laser, both four decades ago, another new contemporary truth became apparent to me. While visually sensitive people would continue to paint and sculpt, these new media required people who were not just visually imaginative but literate, at least literate enough to read a manual explaining how best to use a new technology.

Working long not only in holography but with video and computers, Doris Vila is intelligent, indeed very intelligent in ways that the best artists are nowadays, so different is the art production now from what was done decades ago. (And so different are the best art schools now.)

Since Vila came of age before the art academies wised up, she taught herself mostly, receiving her B.A. degree well after she began to make her high-tech art. For another measure of her considerable smarts, check out her partly eponymous website: Vilamedia.com.

Intelligent artists suffer no doubts about explaining their work, often better than any reviewer can. Read closely what Vila says about her purposes and individual works. Knowing well what she's doing and wants to do, she thus belongs in the great modern tradition of John Cage and Ad Reinhardt, among others, who explained themselves best. Don't dismiss such introductions as "self-promotion," especially when they become prerequisites to better understanding. (Thus have I excused myself from trying to top Vila at explicating individual pieces. Whew.)

Whereas a great painter previously impressed viewers with imagery realized to a level that no one else could equal, now nothing impresses more than looking at great images produced with technology that you can't do—indeed, you can't immediately figure out how it is done, not just because you don't know how but because you lack techno-esthetic imagination.

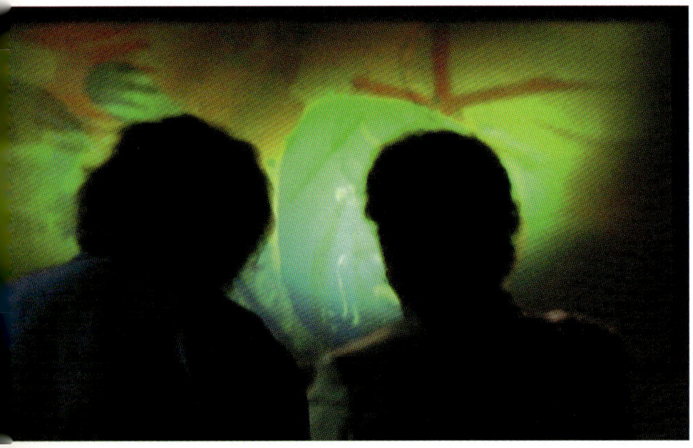

The most common holograms appear on credit cards, the addition of a tiny inscrutable silvery image making the plastic harder to counterfeit, supposedly. Other popular holograms have appeared on jewelry and magazine covers. These are called reflection holograms because the light illuminating them comes from sources in front of them.

In a transmission hologram, by contrast, light coming from behind turns on the image, so to speak. By most measures, this technology encourages richer imagery with more illusion of

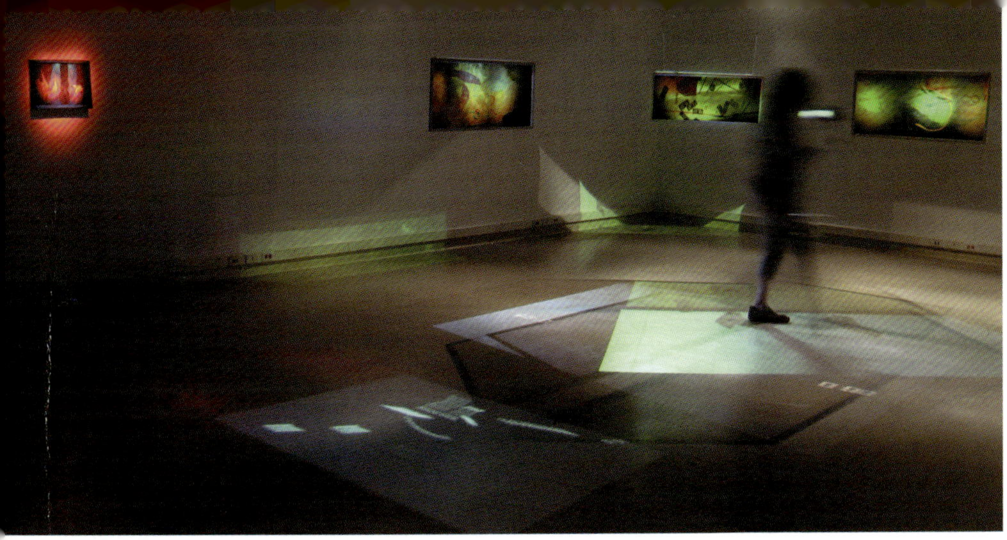

depth than is possible in a reflection hologram. That accounts for why most Vilagrams are transmission in kind (or genre).

One measure distinguishing Vila's holograms from photographs, say, is that they take more time to be viewed. Often, you must move your body to see something initially missed. In this sense, the best holograms are closer to sculpture than photography.

Another mark of Vila's work is size. She thinks big. If the credit card hologram is miniature, so most others rarely exceeded two feet square. She had made displays more than three feet long on one side and has proposed making yet larger holograms.

Because holography has so few practitioners, critical literature about it is slight. One result is misunderstanding about what it is and does. First of all, the produce is Holograms. ("Holographs" are handwritten manuscripts.) Secondly, holography is not just three-dimensional photography.

Four decades ago, I suggested that the best new technological art constituted a true avant-garde, its makers producing works radically unlike anything witnessed before. One theme informing this exhibition is that this thesis still holds true.

Look hard and long at her work to be as impressed as I have been.

Richard Kostelanetz

Individual entries on RICHARD KOSTELANETZ appear in Contemporary Poets, Contemporary Novelists, Postmodern Fiction, Baker's Biographical Dictionary of Musicians, and the Encyclopedia Britannica, among other distinguished directories. For much more, see www.richardkostelanetz.com

Prototype for heaven, home and weightless,
10'x14'x20', screens, light, reflectors, motor, poles, 1988

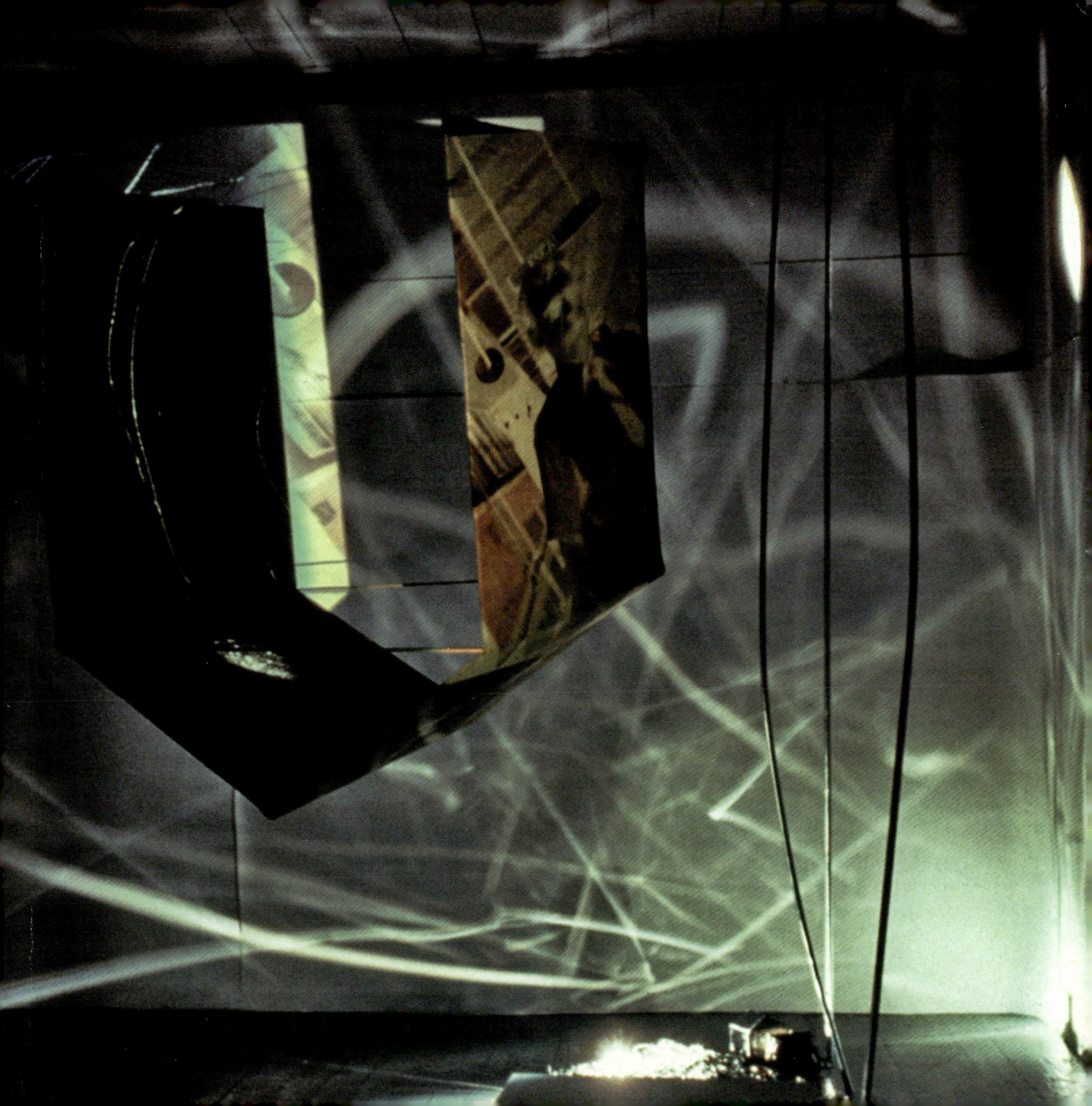

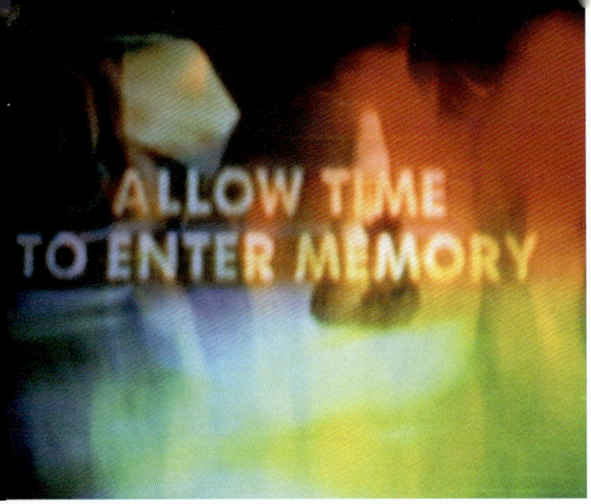

I make holograms of stories instead of things. The picture at left is a still from **Ideal Fabricator**, where a projected video signal reconstructs the holographic image. This piece is part of an ongoing series called **Prototypes for a City Machine**, where elusive forces for personal and social change are gathered into post-scientific inventions.

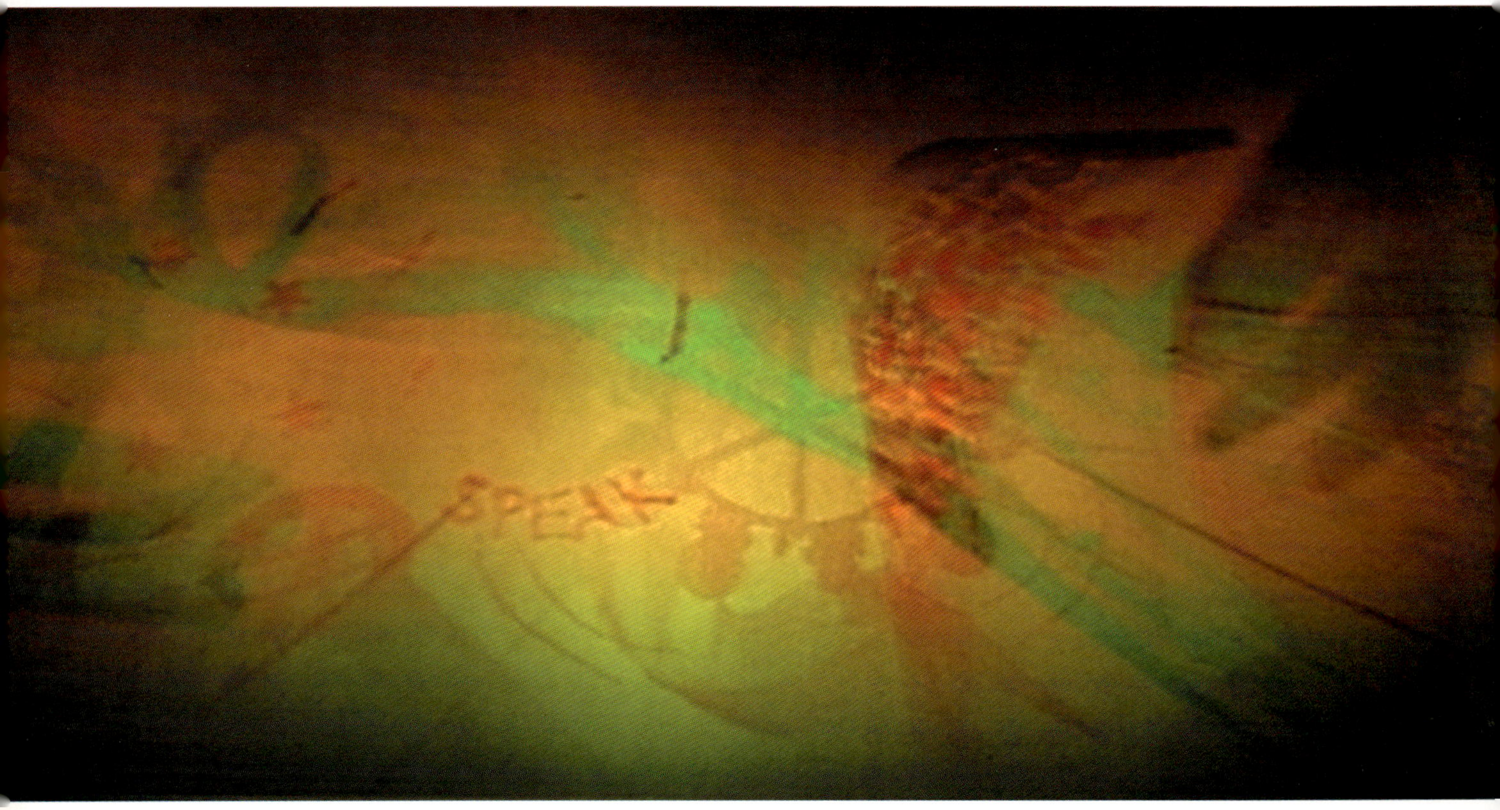

Decision corrector (with short-range-past capability), white-light (WLT) hologram, 21"x42", 1992

The first studies for this series became five large holograms in 1992, entitled
a positive destiny accelerator,
the decision corrector (with short-range-past capability),
a flash-driven media-magnet,
a valve for the distributor of worth and
a brilliance enhancer.

In traditional holography, it is common practice to look for deep stationary objects to record. Instead, I see the holographic environment as an arena for viewer discovery and participation. Building on the ideas of Moholy-Nagy's time-based work with light, these pieces expand a visual dialogue out in space and time.

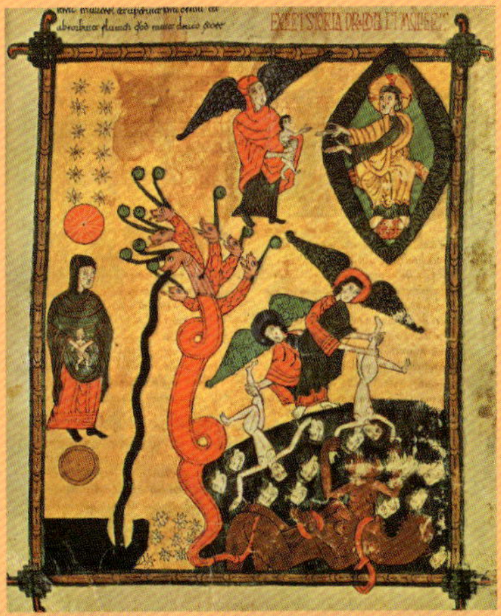

Like pre-Renaissance medieval art, this is a narrative approach which is non-linear. Before the science of perspective was adopted, a single picture could depict several moments in time. Space was often arranged, not for topographical correctness, but to show hierarchies of meaning.

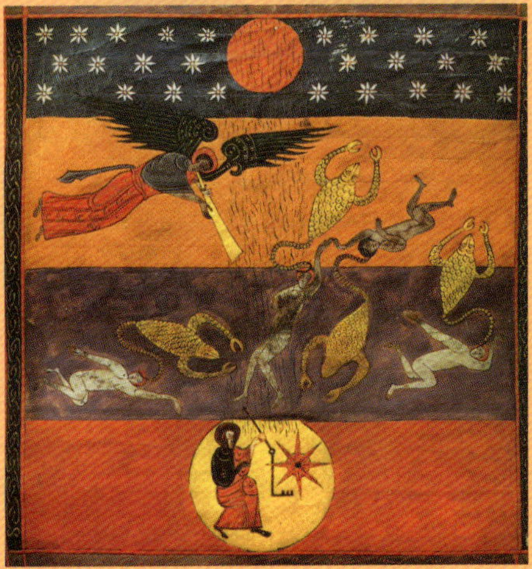

A fine example is the Beatus of Liebana, a series of illuminated manuscripts of the Apocalypse from 10th century Moslem-ruled Spain. Its rich color, patterning and mystical allegories still attract the attention of scholars. Umberto Eco said "...when the language becomes able to present something to our eyes as if we could actually see it...to represent not a still scene but something in motion, something that could be visually translated only by a movie...Mozarabic illuminators were obliged to represent it by a still and two-dimensional image...."

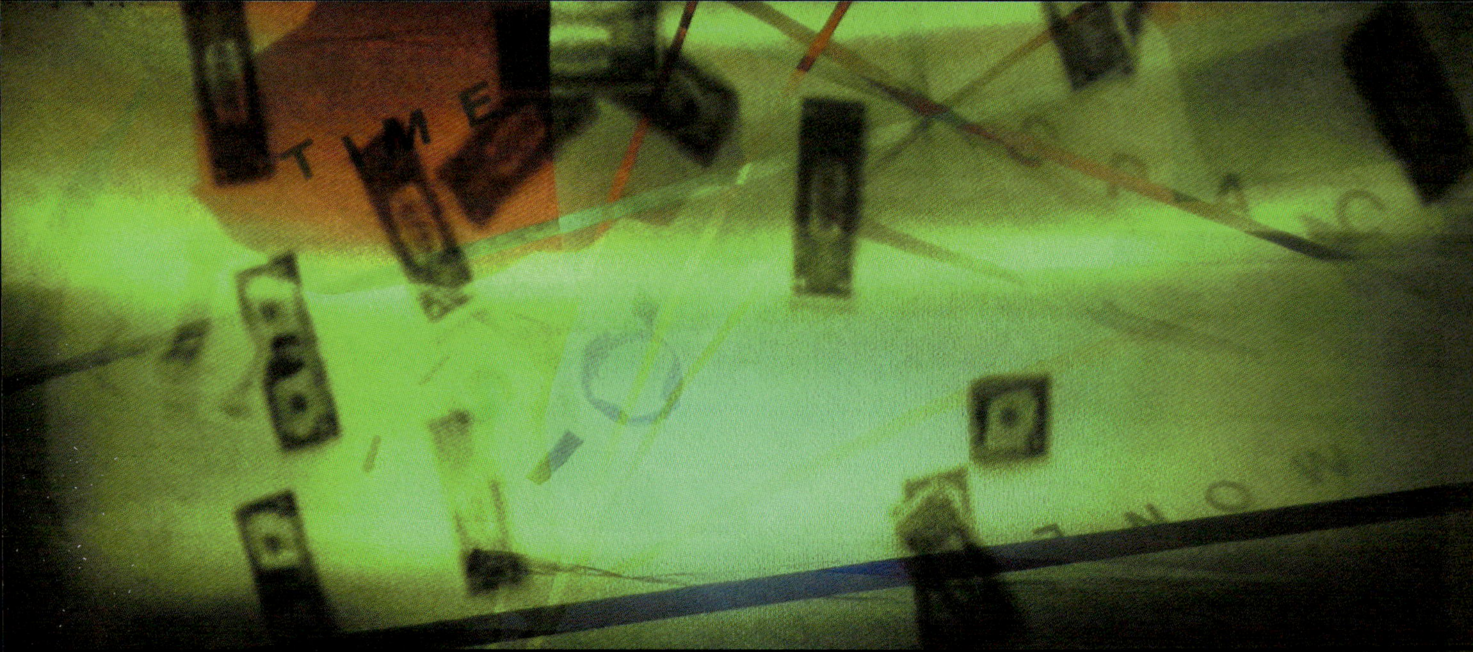
TIME
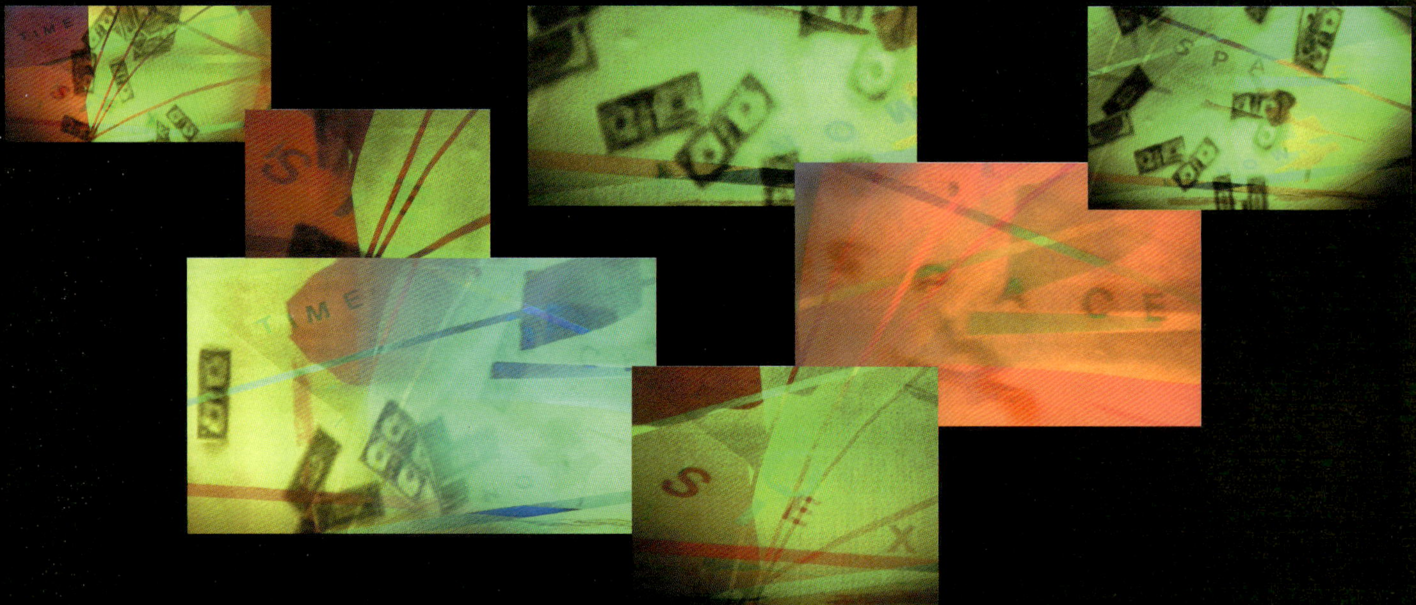
TIME

Yet, the three dimensions of holographic recording make me crave the presence of the fourth dimension – time. Since slide projector experiments in 1983, I have been reconstructing holograms with projected images. As the first to use changing projections this way, for clarity I dubbed them 'prolopics' (for projection-holo-pictures). In 1988 **Chart: space-time-sex-money continuum** was shown reconstructed by an 8mm film loop of falling dollar bills. The film imagery was caught and focused on the hologram plane, while the cloud of holographic words and chart graphics floated three feet in front.

In 1989 this led to the installation **Overview of a free-guilt economy**, a diptych composed of two holographic panels **He Said** and **She Said** facing each other about nine feet apart. **He Said** is lit by a 16mm film loop of fire and **She Said** has a projection of a 16mm water whirlpool with regular streaks of red. The light from the two films mixes on the floor between the holograms' while the observer is caught in the visual ping-pong of a conversation between the two halves.

With the availability of video projectors, the moving image now shines through a single lens, ideal for illuminating and reconstructing holographic images. At last, a live image can be projected and a bridge links holography with electronic imaging for public spaces.

Doris Vila

Chart: space-time-sex-money continuum,
17"x39", WLT hologram, shown with varying views, 1988.

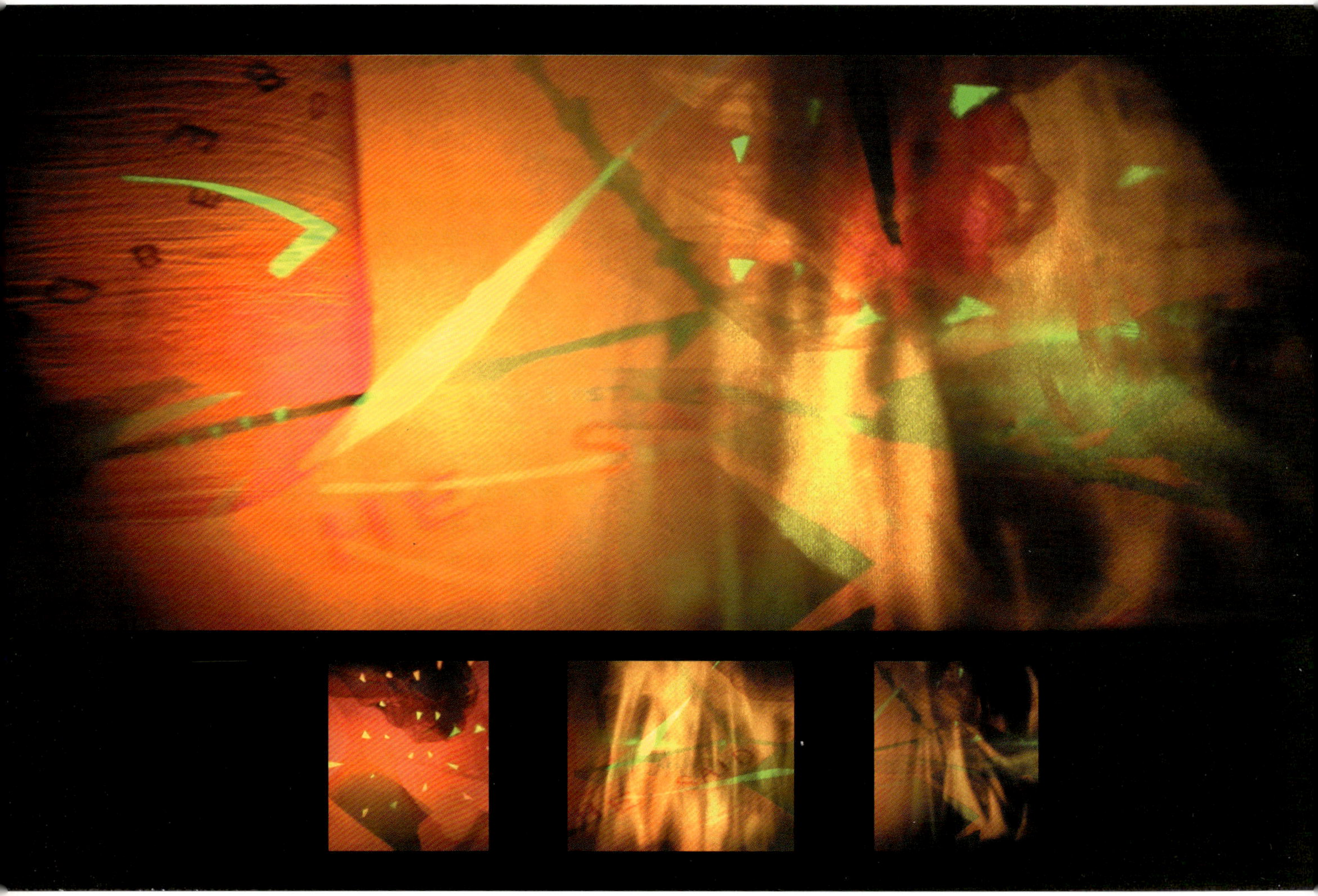

He said

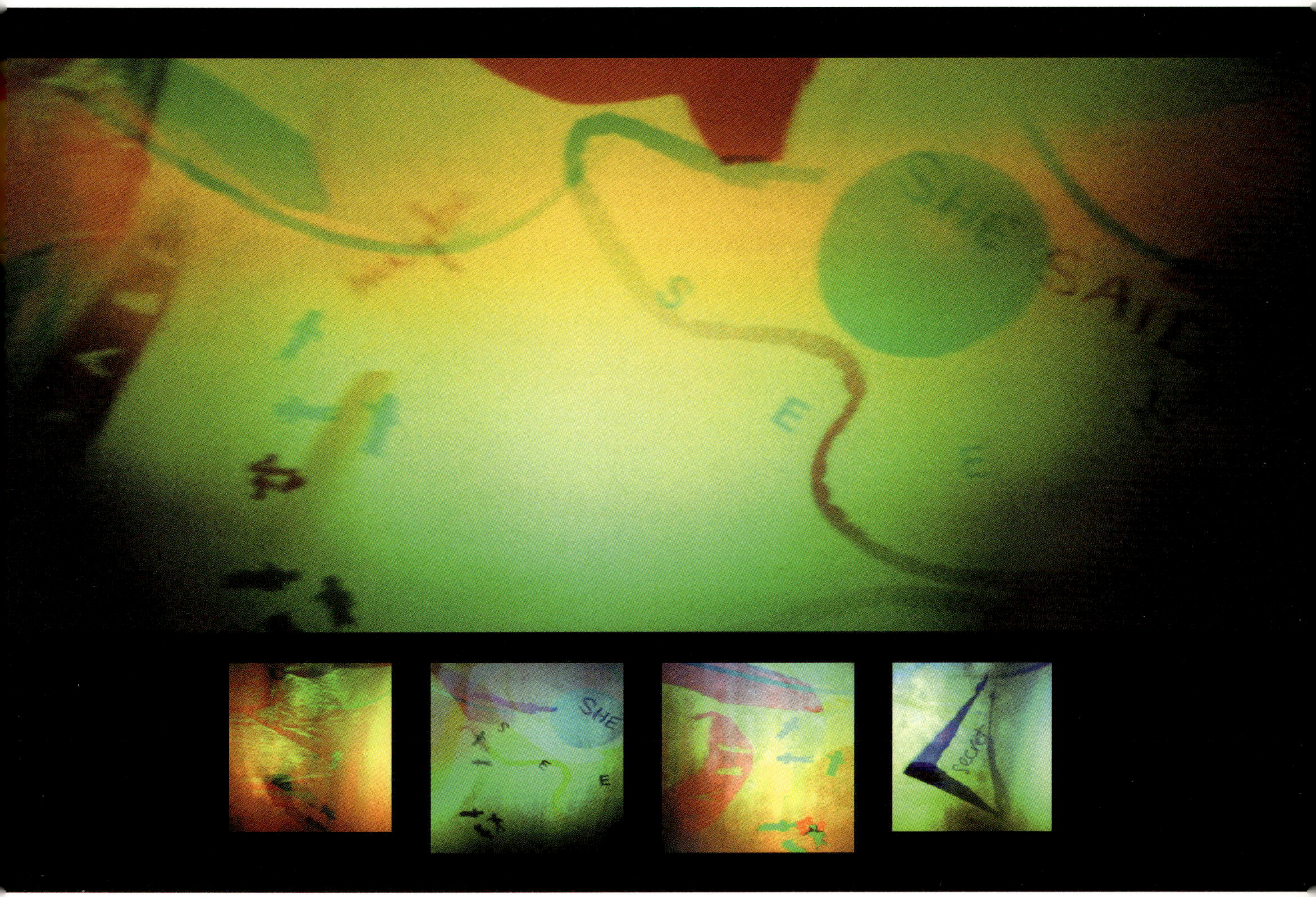

Overview of a free-guilt economy, two facing panels **He said** and **She said** (look for the secret), where the observer-child is caught in the crosstalk of the parental conversation, seeing/hearing one at a time, while the other parent is behind, standing where projections of fire and water intersect on the floor, installation with two WLT holograms, each 20"x42", 1991

She said

When light falls,
80"x27"x36", holographic installation with light scanning from top to bottom, continuously falling, 5-second cycle seen in the three small photos at right, 1983.

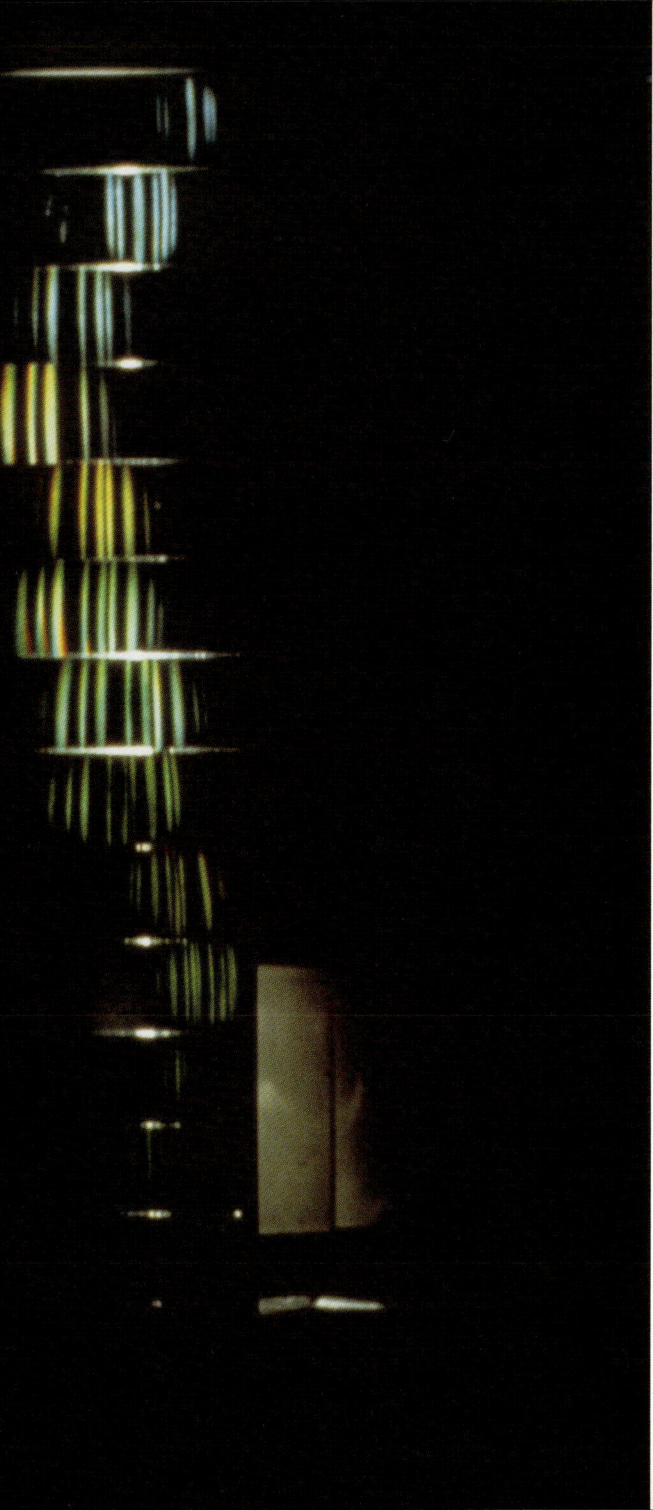
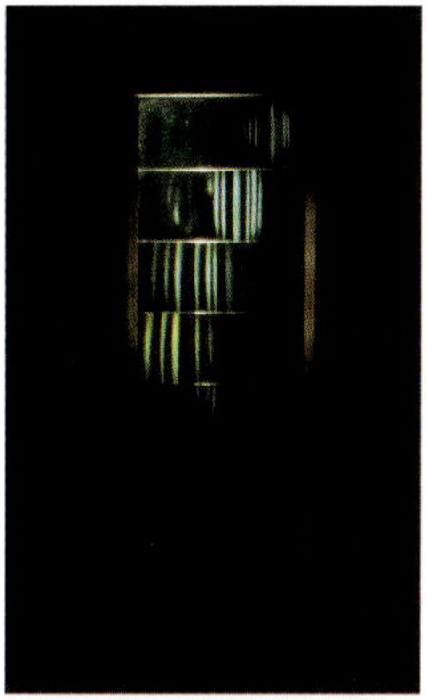
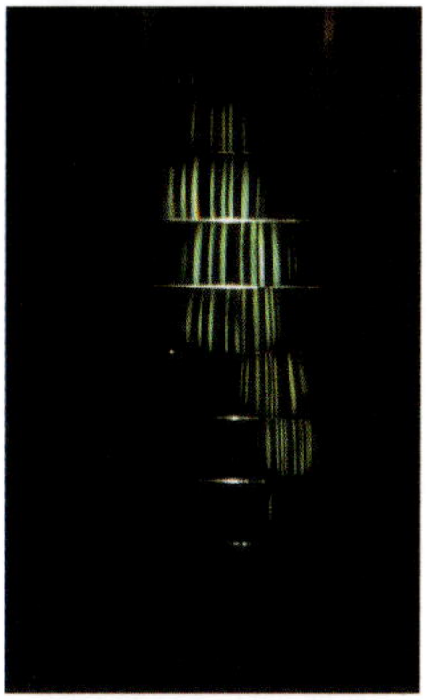
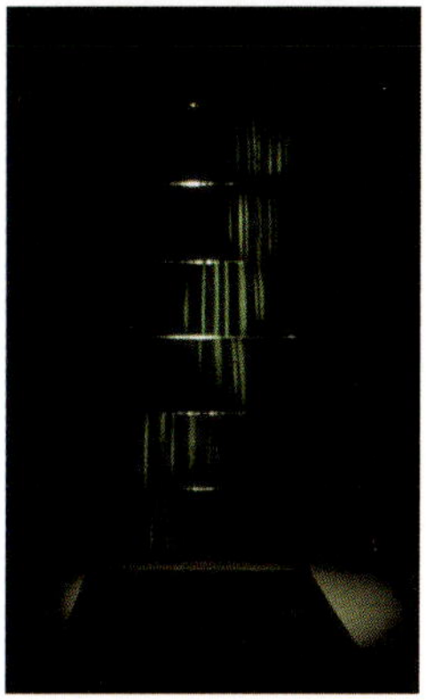

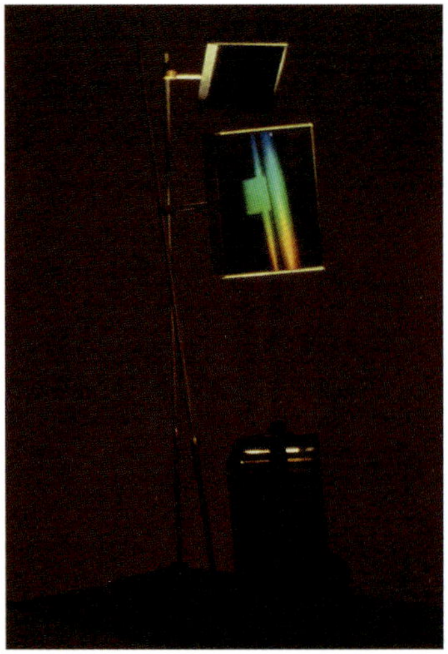

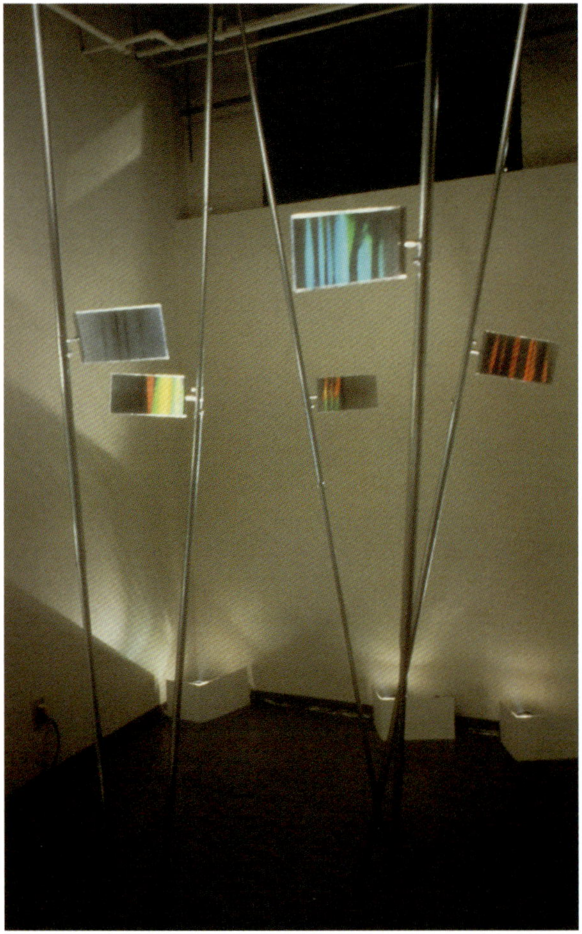

Flag (above) poles, hologram, mirror, 18"x9"x9", 1984

Meanwhile (right) installation with telescoping metal poles, floor-to-ceiling, 5 WLT holograms, 12'x6'x6', Toronto, 1985

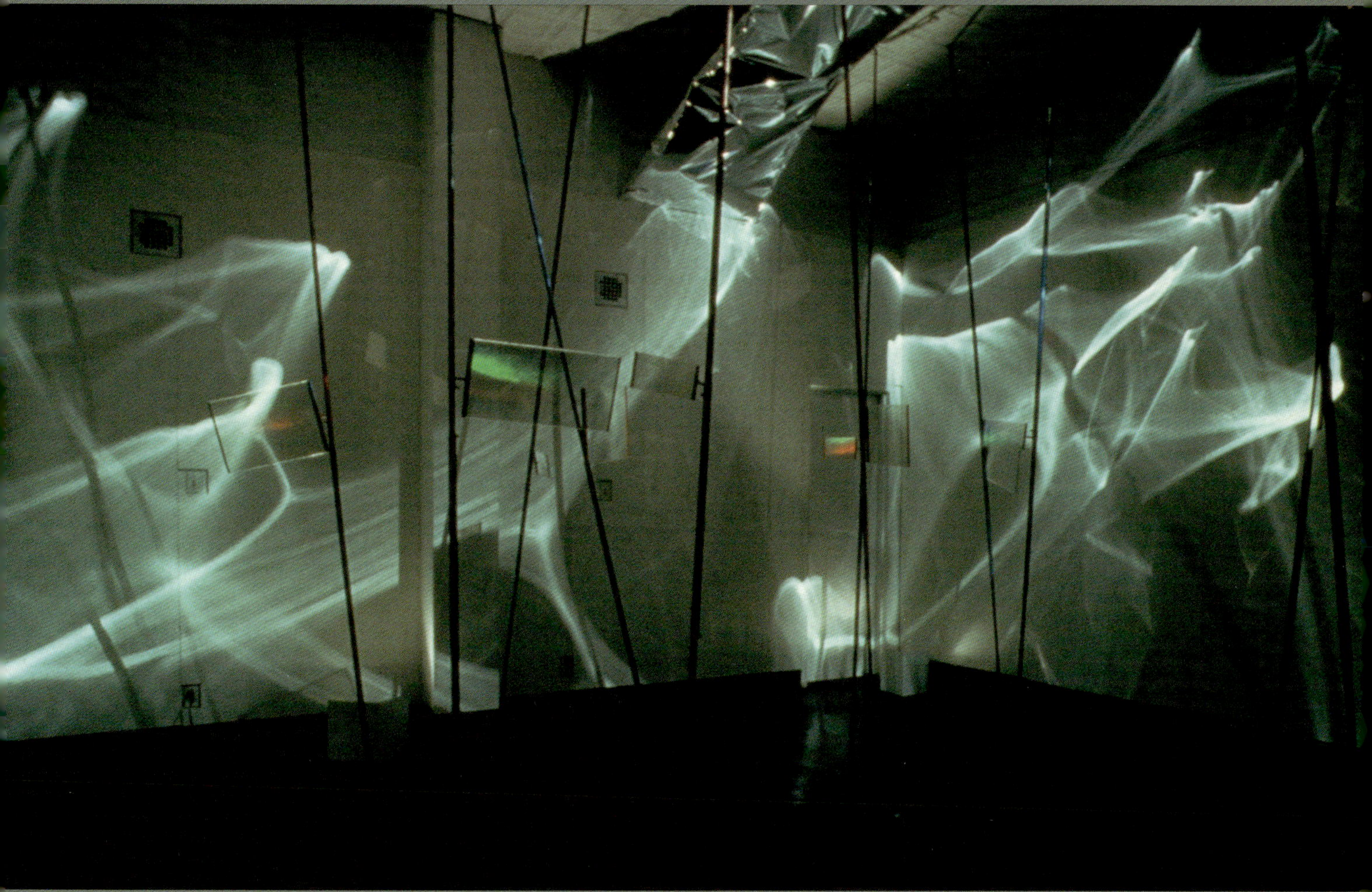

After many days had passed

Full-room installation, 13'x15'x18', at BACA, Brooklyn, New York. Viewers walk among the metal poles with holograms mounted leaf-like on the side, while a continuously repeating lumia pattern washes over the walls. 12 white-light transmission (WLT) holograms, 3 lights, mylar arch, projector, motor & 10-second light scan, 1985.

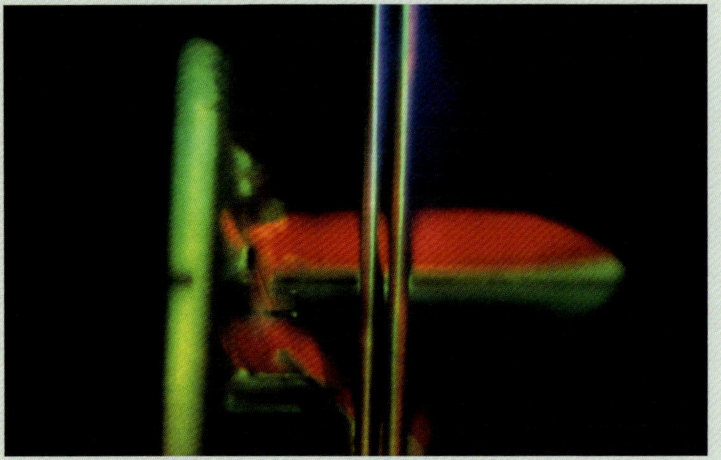

Chiasma,
WLT hologram,
5" x 8", 1981

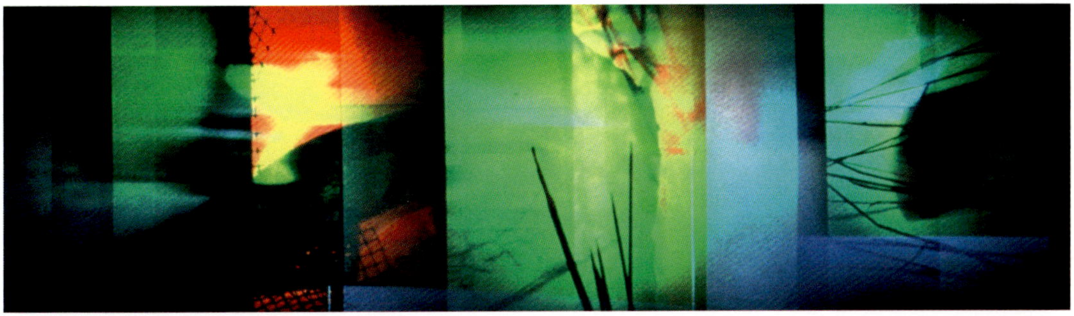

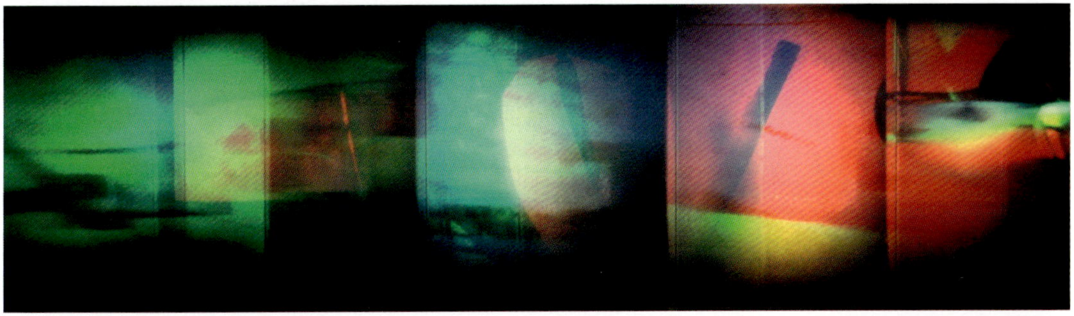

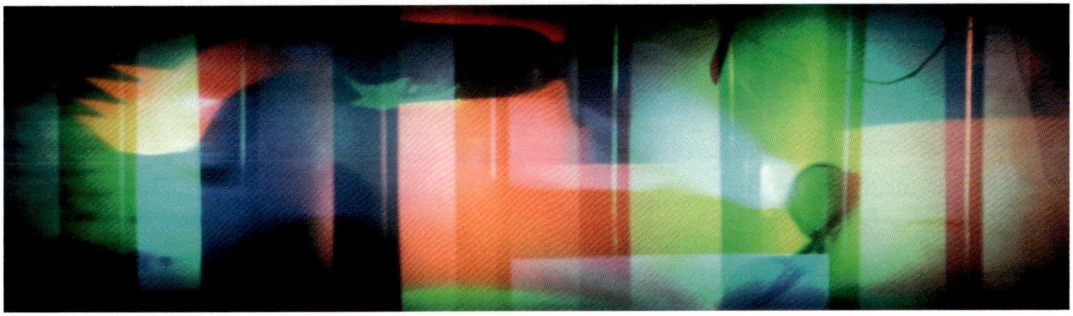

Spatial phrases: WLT holograms, shadowgram, photogram, found object and stenciled imagery mix in layers of color with sixteen exposures, 1985. The long strips catch a phrase and involve the viewer in a discovery sequence invoking the privacy of the act of perception. Starting from the top:

Exactly as we remembered, 10"x51"
Heading east on noise, 10"x50"
Just a figure of speech, 10"x50"

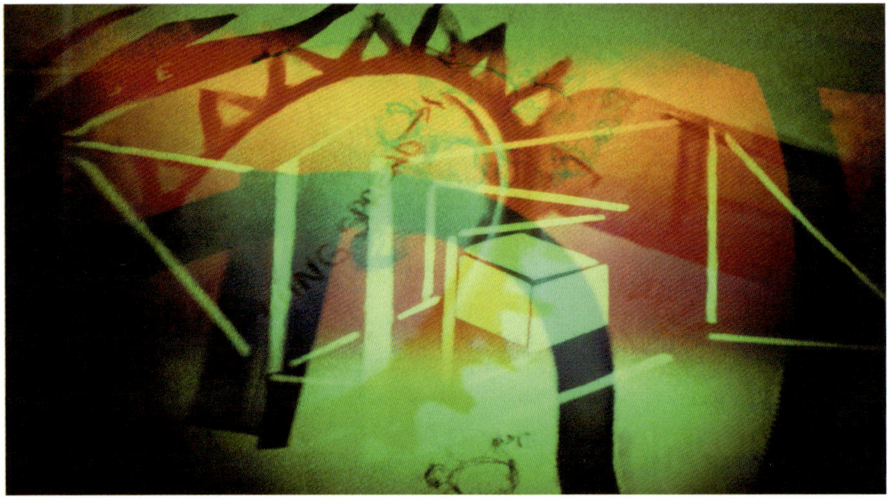

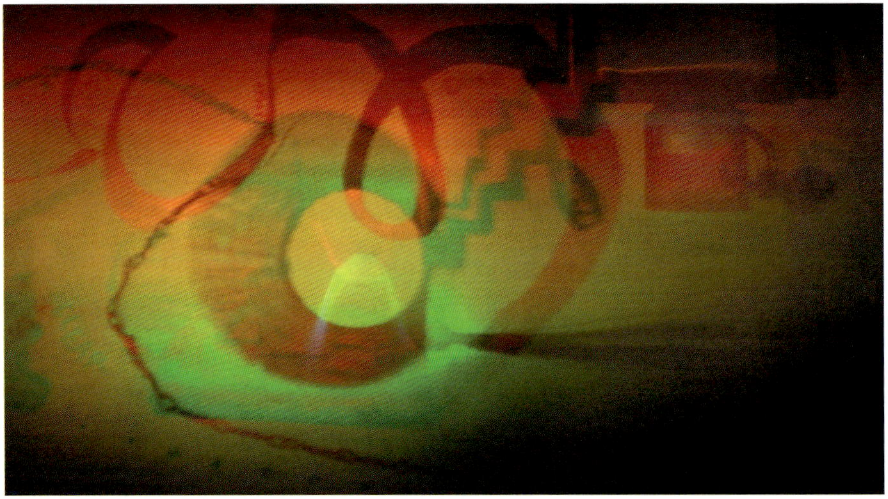

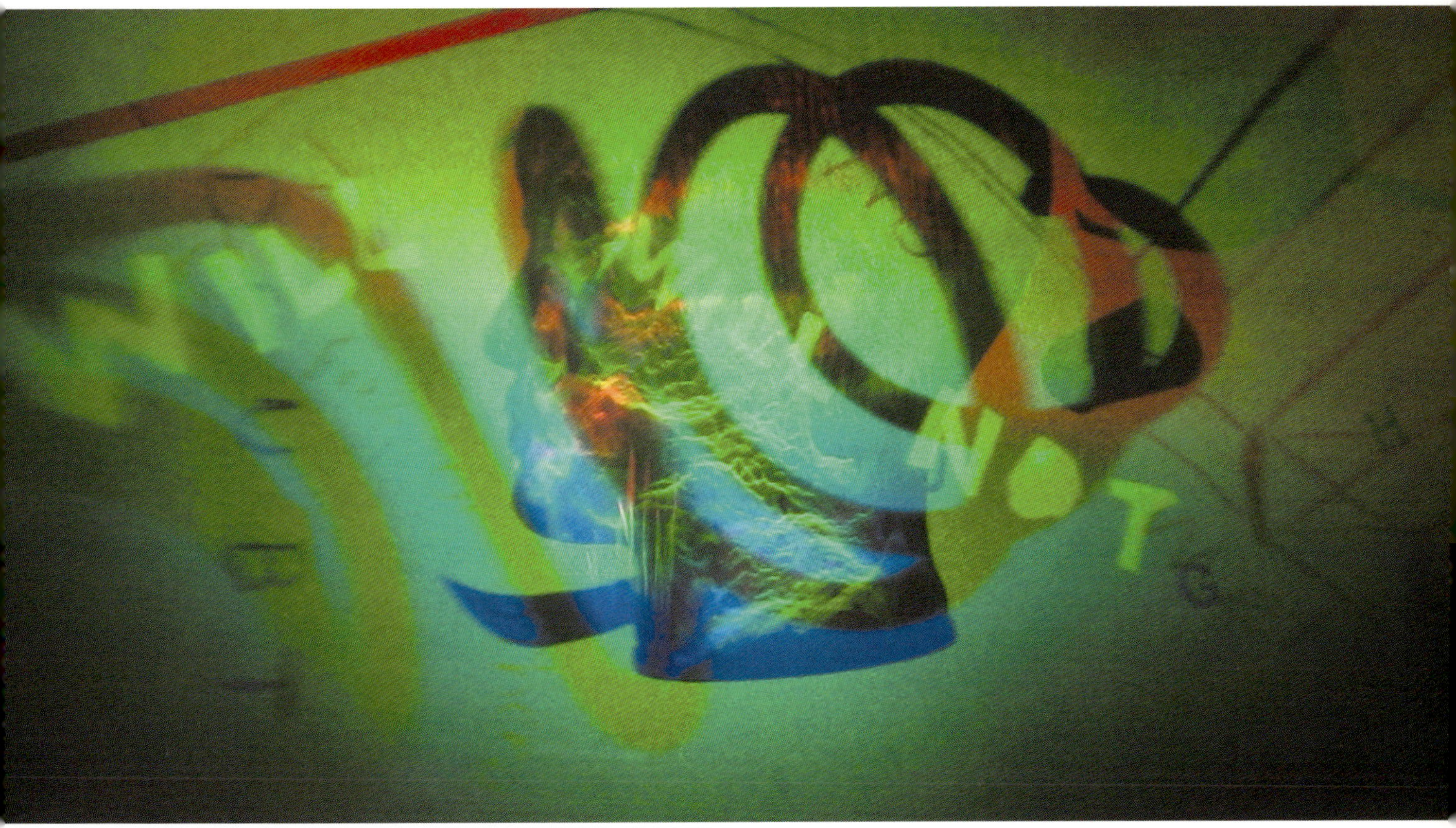

explication series: charts, graphs and maps of pseudo-scientific paradigms rendered in large multi-exposure WLT holograms, with imagery projecting up to 4 1/2' in front of the film plane.
Diagnosis of an existential headache, 17"x40", 1987 (above)
Blueprint for a metanatural power plant, 18"x43", 1988 (at upper left)
Easy rollover power broker, 20"x36", 1992 (at lower left)

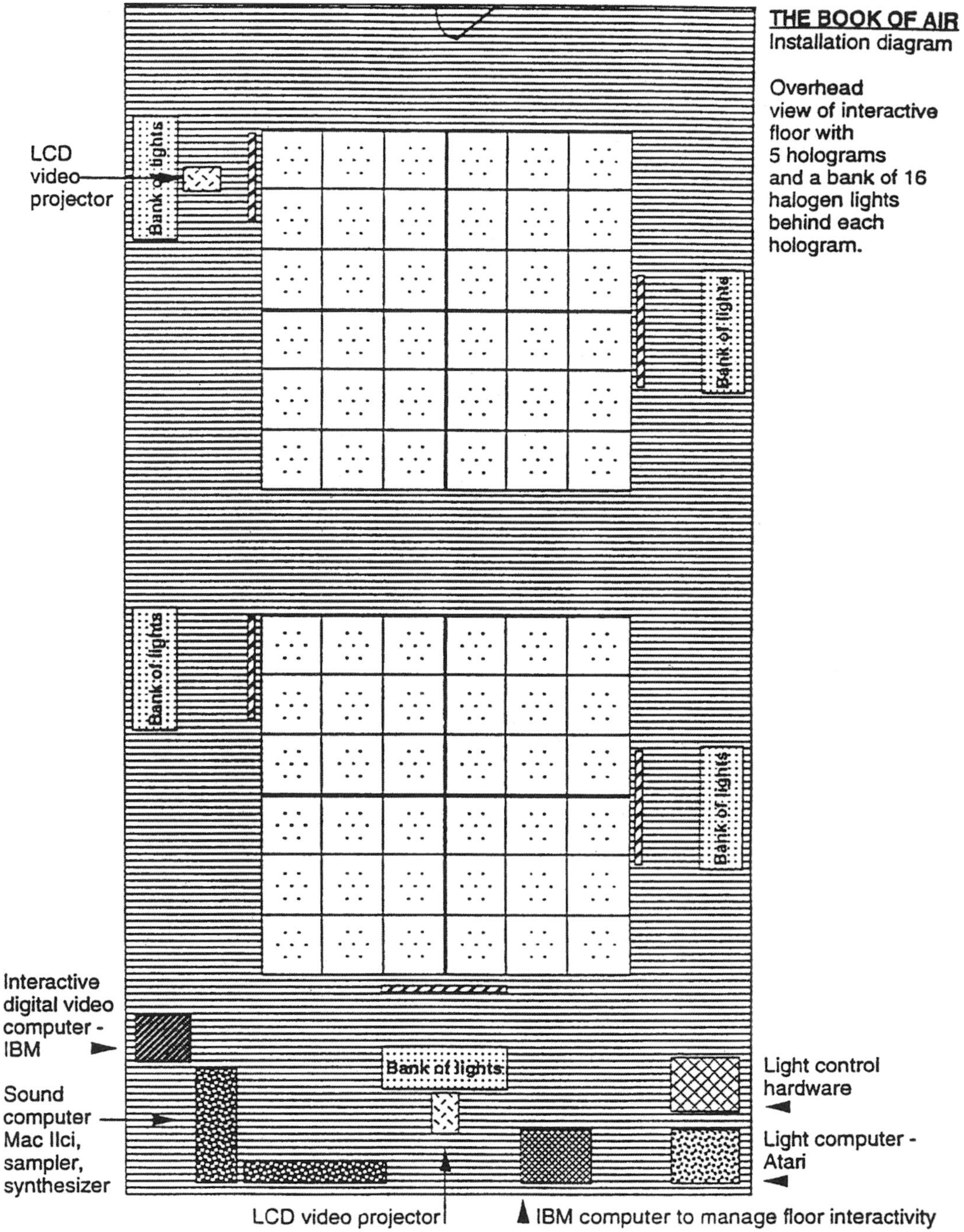
THE BOOK OF AIR
Installation diagram
Overhead view of interactive floor with 5 holograms and a bank of 16 halogen lights behind each hologram.
LCD video projector
Bank of lights
Bank of lights
Bank of lights
Bank of lights
Bank of lights
Interactive digital video computer - IBM
Sound computer Mac IIci, sampler, synthesizer
Light control hardware
Light computer - Atari
LCD video projector
IBM computer to manage floor interactivity

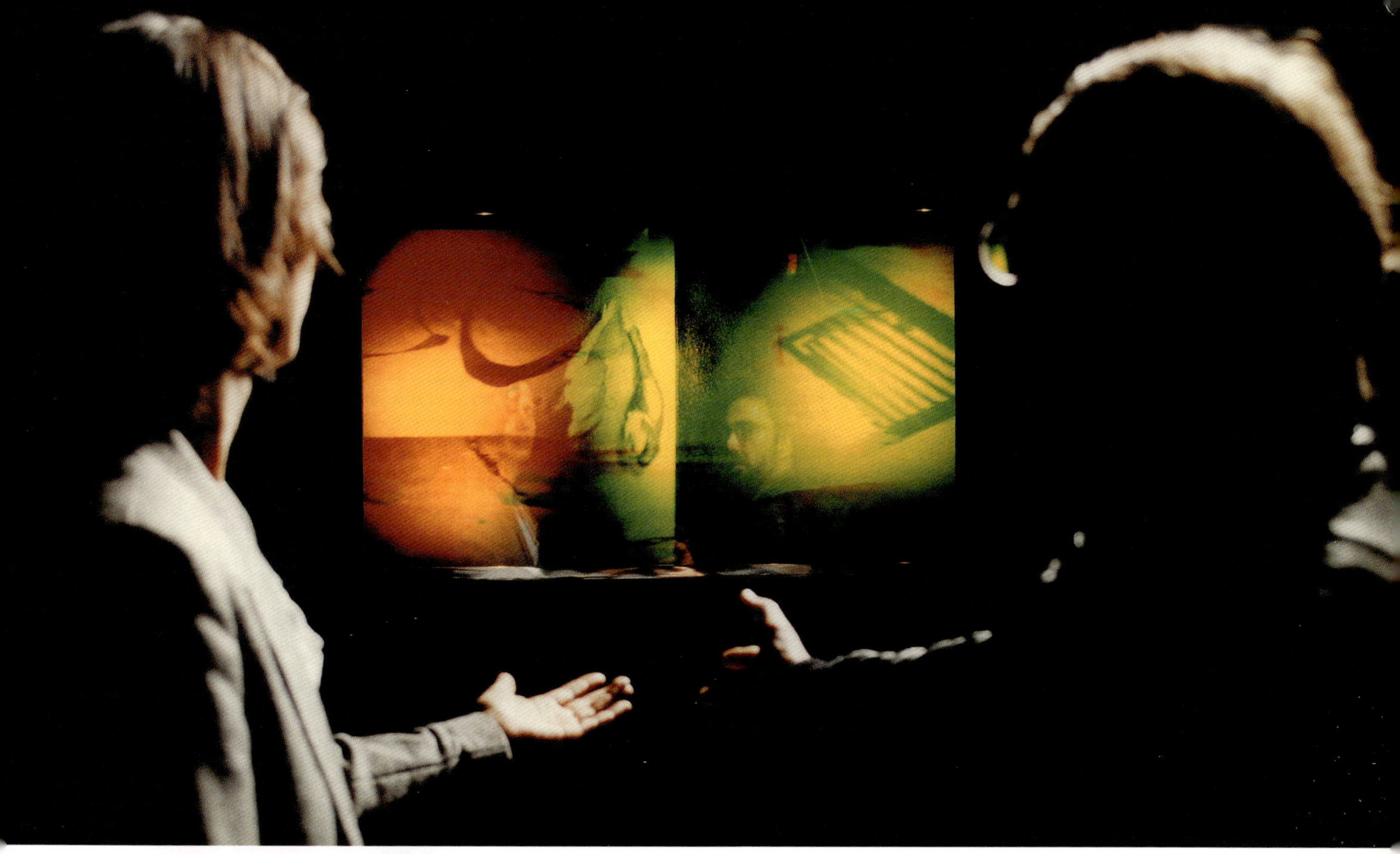

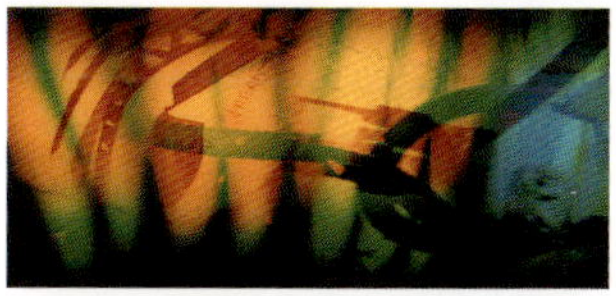

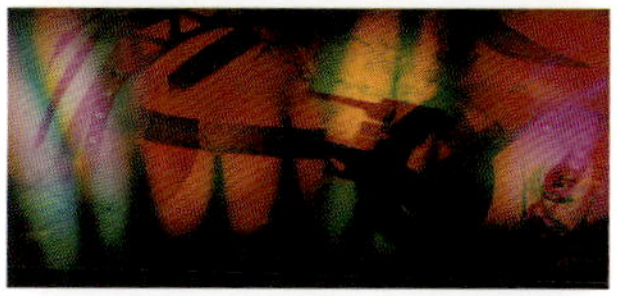

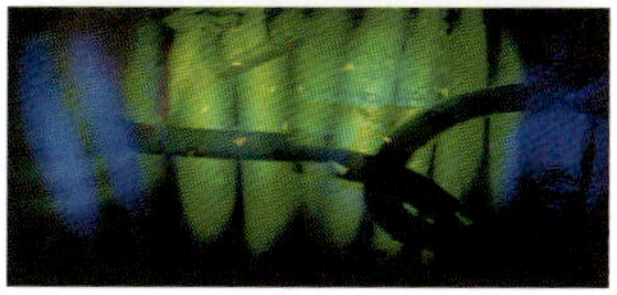

The Book of Air, interactive installation at Chip Arts Atelier, Bonn, Germany, 16'x20'x50', 1993. Viewers enter a room with 5 large holograms. As they walk across the floor, their infrared shadows trigger computer-controlled sound, video and lights which in turn animate the holographic imagery.

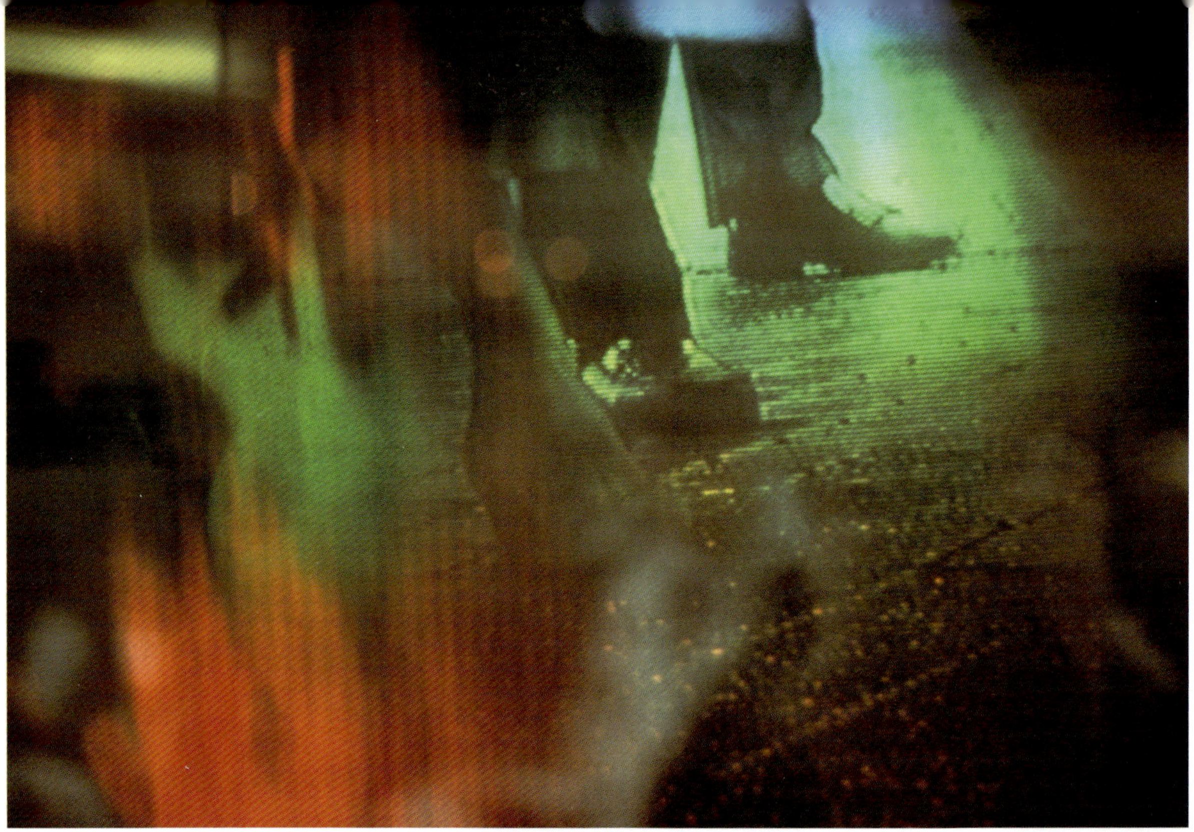

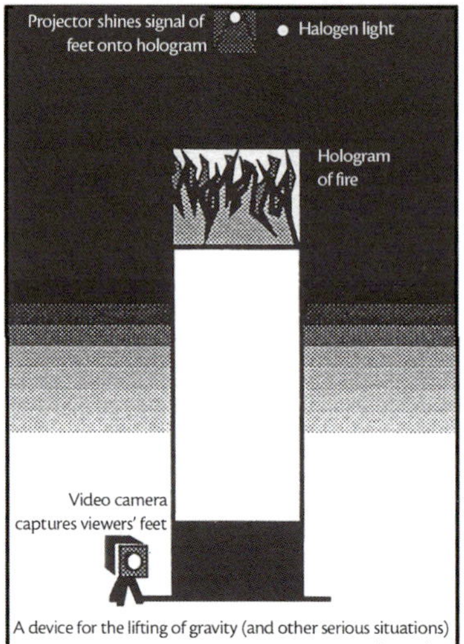

A device for the lifting of gravity (and other serious situations)
Front view

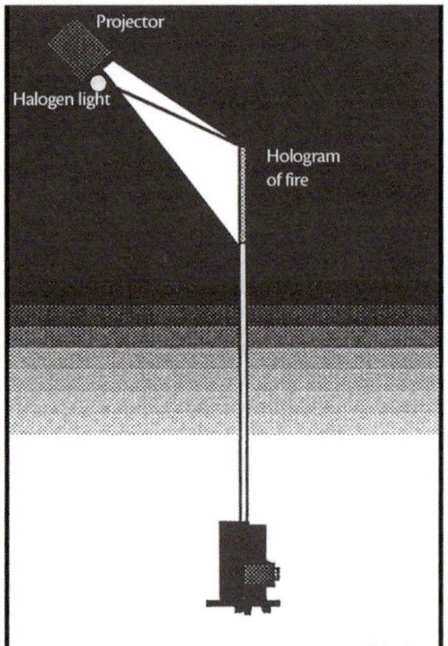

Side view

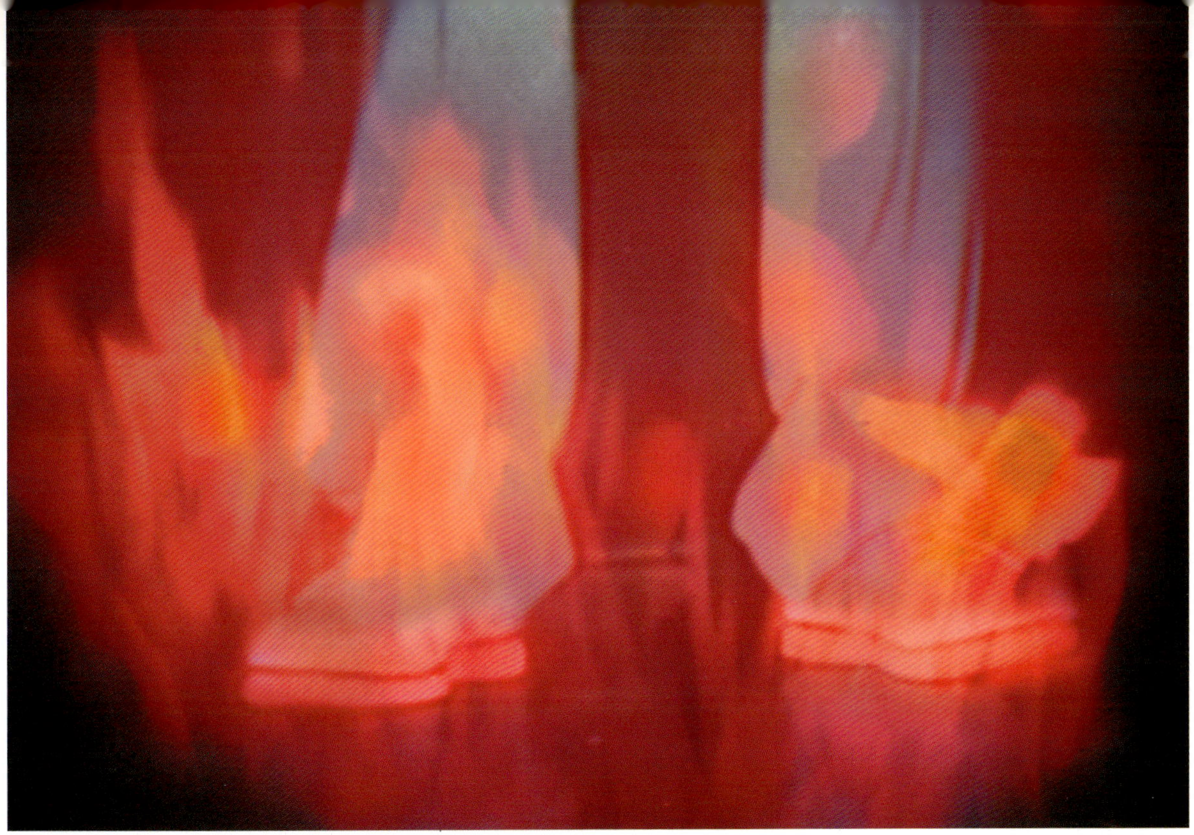

A device for the lifting of gravity and other serious situations, at left, installation in a store window, 7'x3'x3'. A hologram of fire is reconstructed with live video of the feet of passersby on the main pedestrian shopping street of Cologne, Germany, 1992

A device for the lifting of gravity (again), above, on a reflection hologram of fire, 2008

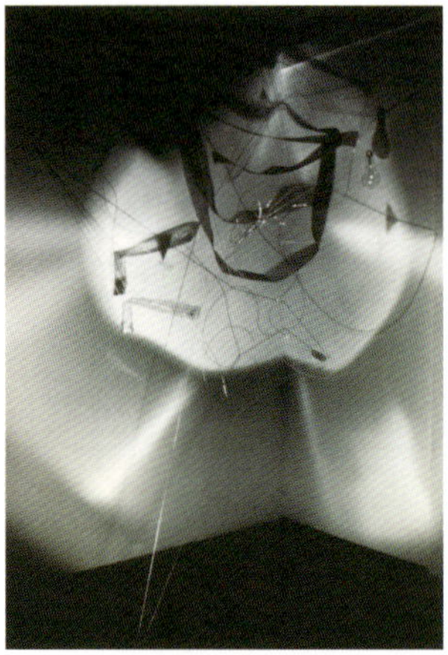

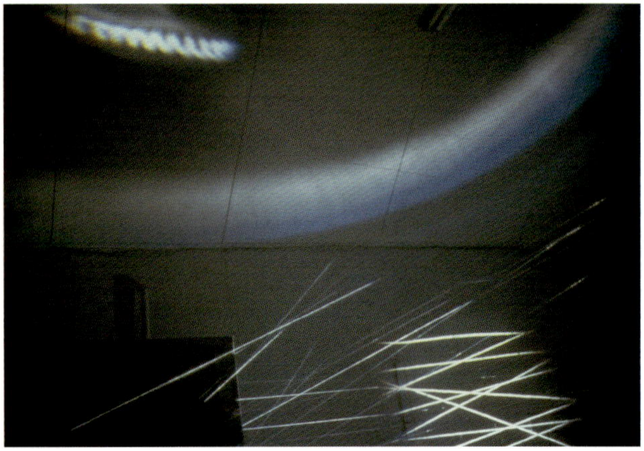

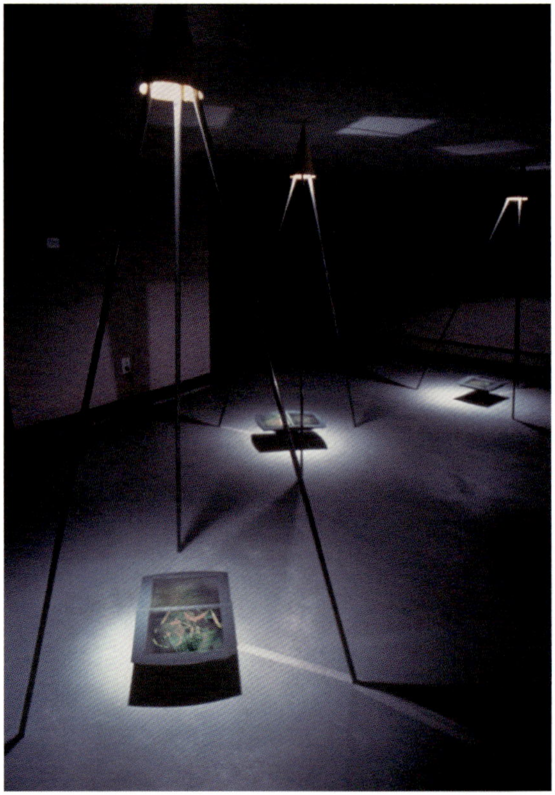

One-point landing, corner piece, 5'x5'x8', 1985

Relativity joke, full-room installation, 16'x18'x20', 1986. Projected lines orbit the room crossing strung nylon lines overhead, so points of light appear to travel at many speeds at the same time.

em-creatures (empirical emotional phenomena), 76"x12'x20', installation with 3 tripods, each with light and 2 revolving reflection holograms, 75"x39"x39", New York, 1990.

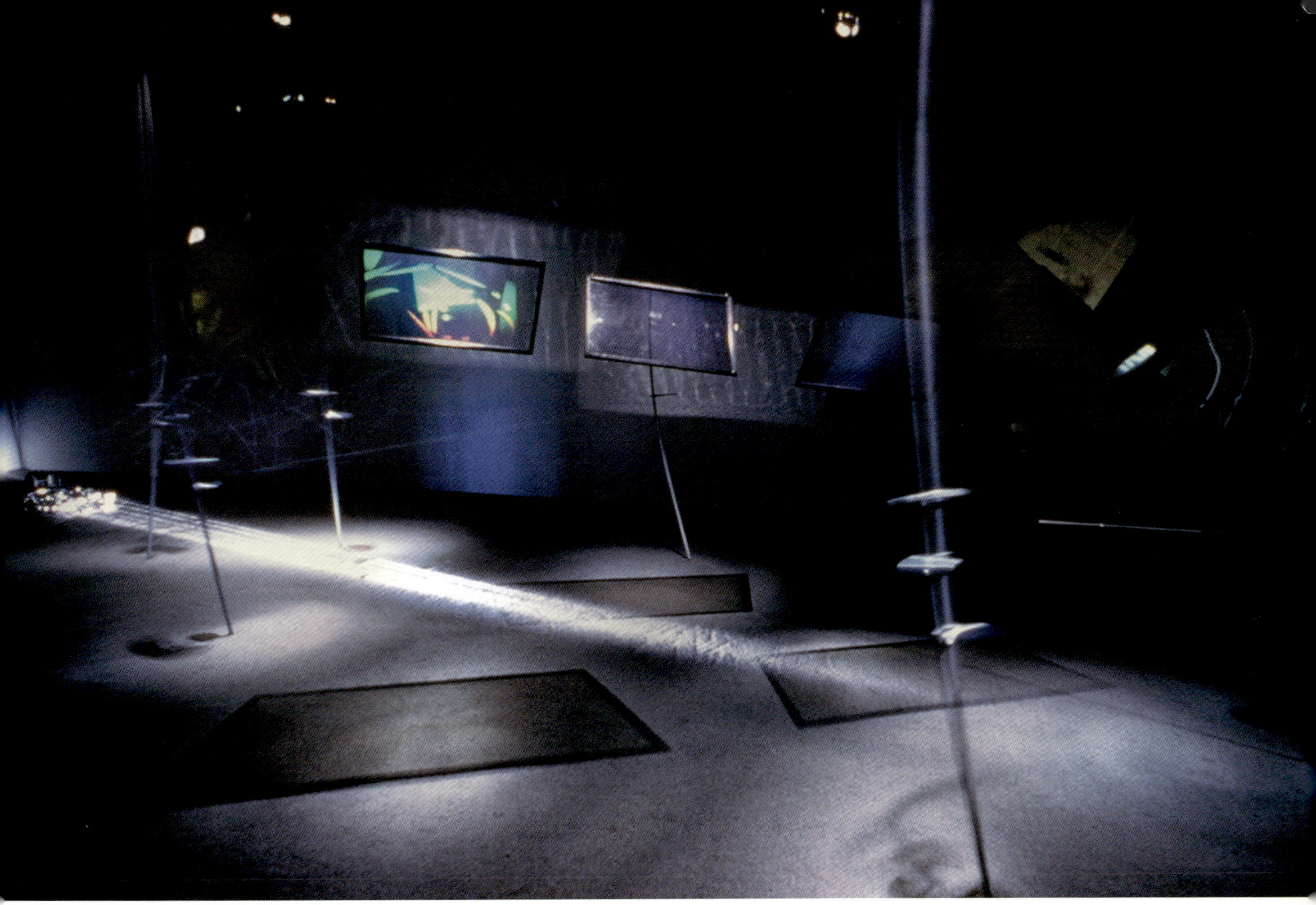

Heaven, home and weightless, installation at Photographic Resource Center, Boston, 13'x40'x28', 1989. Viewers enter a large room inhabited by slowing turning poles carrying semi-circular reflection holograms. A large screen calmly opens and closes, with a projected image sliding across it. Three large graphic transmission holograms and two film loops come into view depending on the viewers' position. A continuously rotating patch of light, the viewers' presence and shadows, and 2 spatial sound loops complete the piece.

Wave Hill, one of New York's most beautiful gardens, commissioned a light installation for their classic greenhouse. The entire conservatory became a natural clock, suggesting the rising sun in the east to the flickering filaments of the twilight in the west.

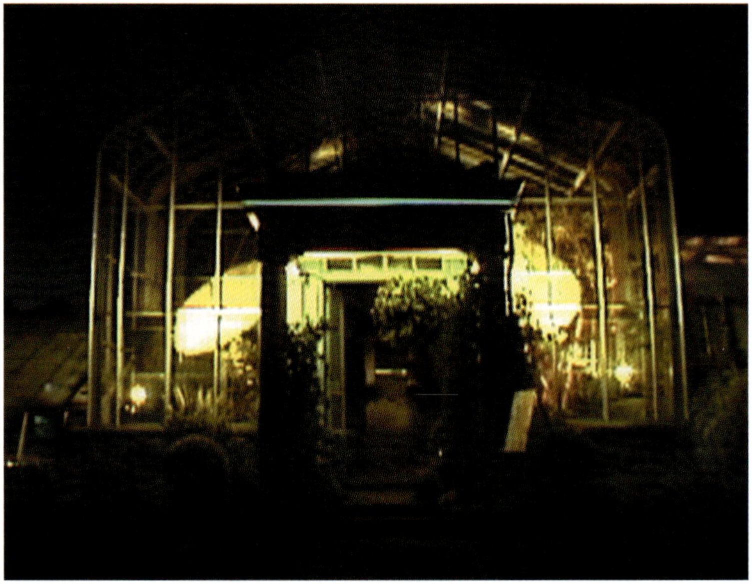

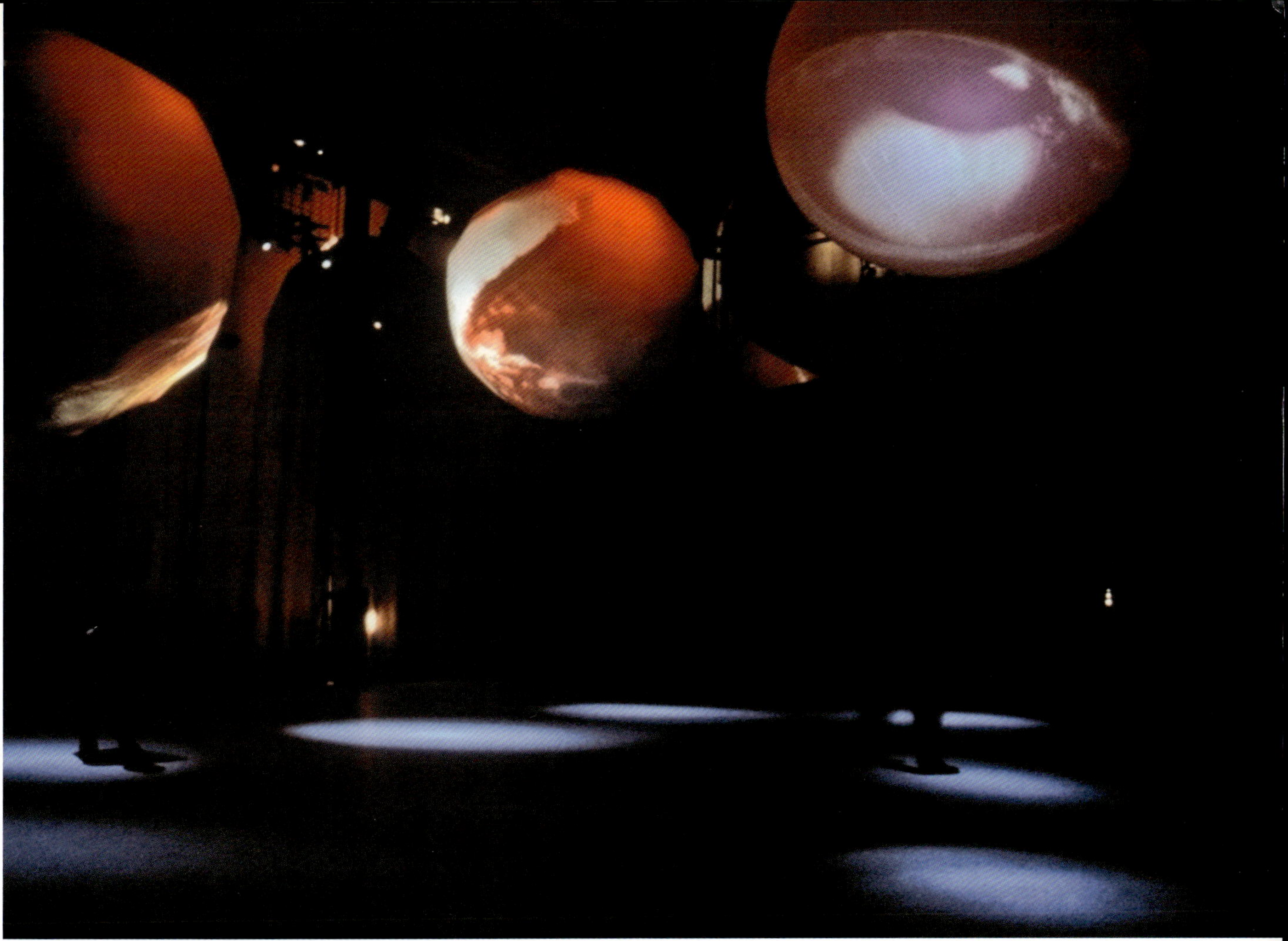

Theatermachine, collaboration with BEC, Germany, experimental media stage. In a large room with 20′ tall ceilings, visitors glide through zones of a responsive arena, triggering short sound fragments from high points of opera and film. Just as you recognize the theme, it disappears. Visitors' motion also triggers video projection on three 8′-diameter spherical screens floating overhead. On the wall, flocks of eyes and herds of words swarm to the visitor via real-time artifical-life animation. In six areas of the room, people's motion sparks a sound-bite on one of these themes:
the zone of the unexpected, meeting and innocence; the conflict camp;
the magnetic field for hope and found love; the region of death;
the area of despair, lost love and suspicion; the house of will and victory.
Tapetenfabrik, Bonn, Germany, 1995.

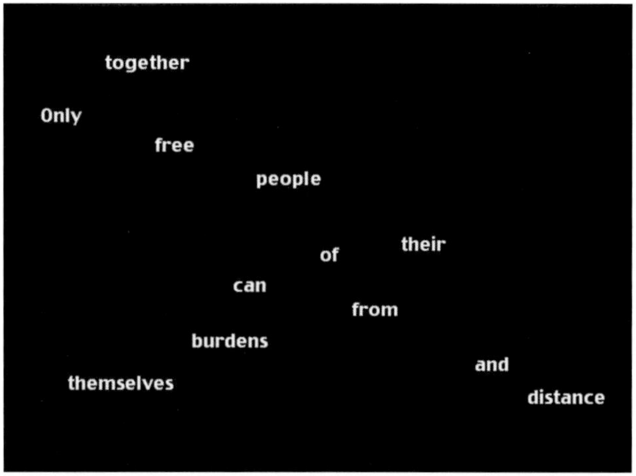

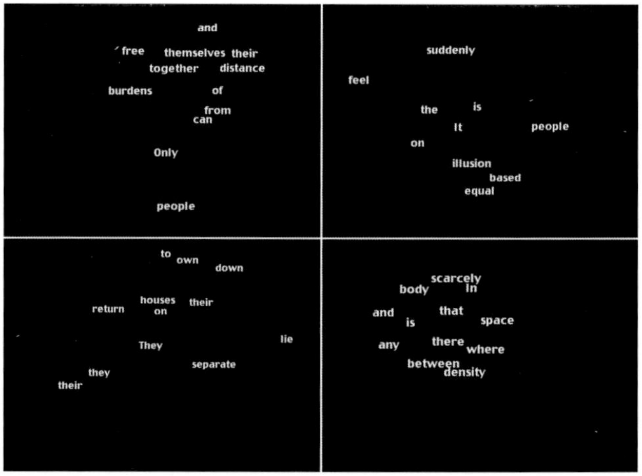

A Flock of Words, a cross-media collaboration with composer Robert Rowe and Eric Singer, peformed at NYU. This work revolves around groups in motion – flocks, swarms, and crowds. A text from Elias Canetti's **Crowds and Power** flies in and out of readability through a swarm of birds. Computer signals trigger light and sound events, including light sweeps, video and words animated by "woids", an artificial life algorithm based on flocking and herd behavior.

Only together can people free themselves
from their burdens of distance; and this, precisely,
is what happens in a crowd.

It is for the sake of this blessed moment,
when no one is greater or better than anot
that people become a crowd.

It is based on an illusion;
the people who suddenly feel equal
have not really become equal;
nor will they feel equal for ever.
They return to their separate houses,
they lie down on their own beds,
they keep their possessions and
their names.

Speaking volumes, reflection hologram, 16"x12", 2001

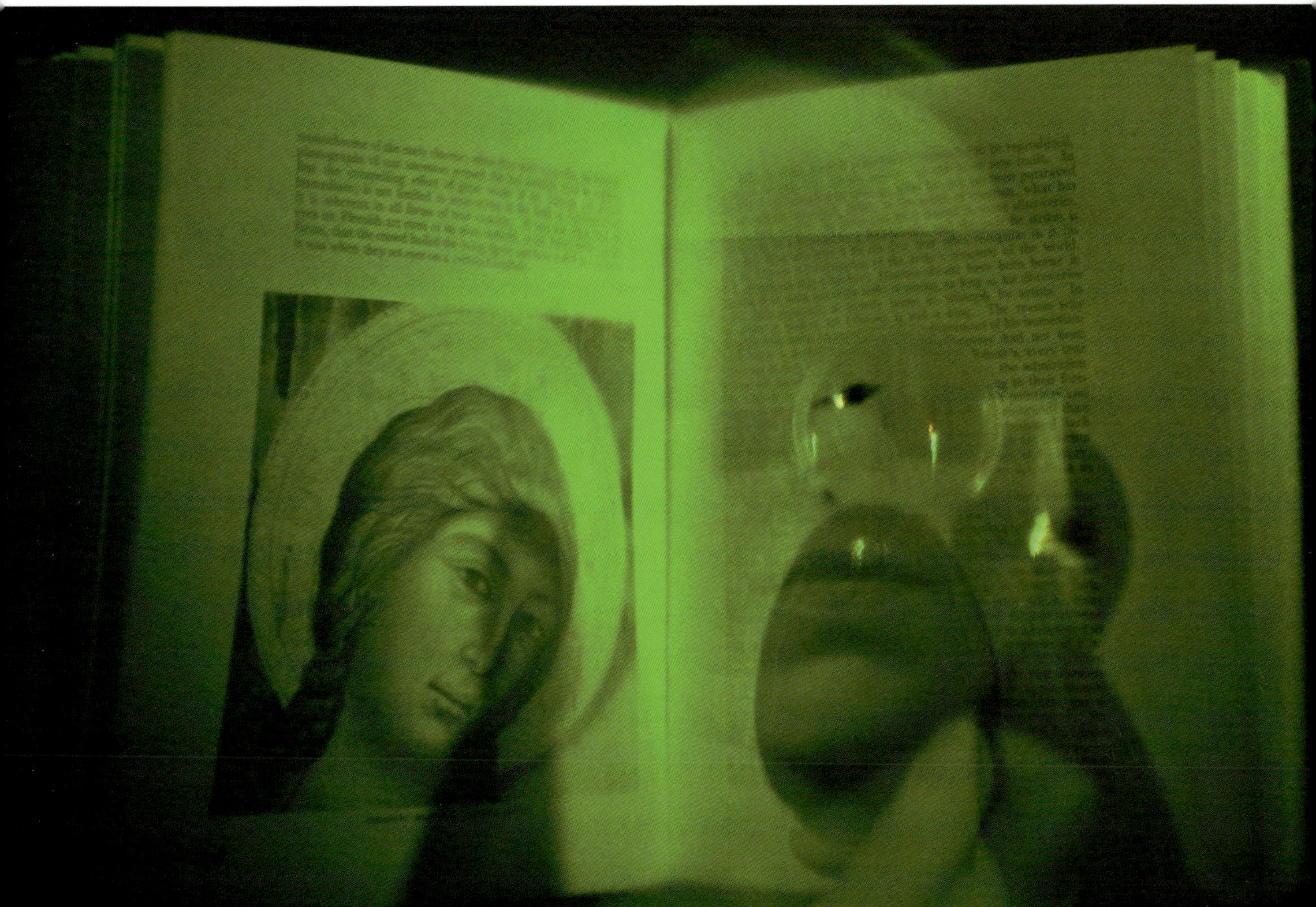

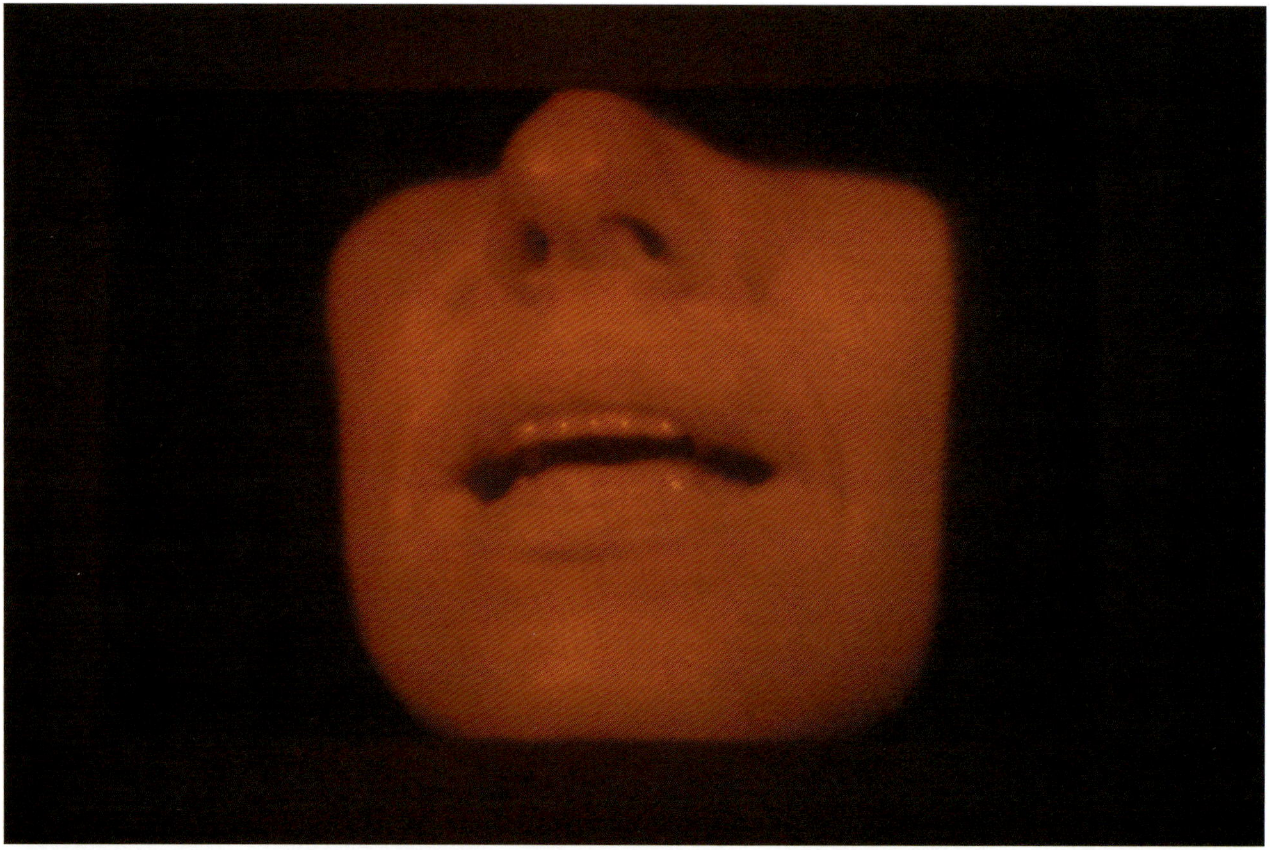
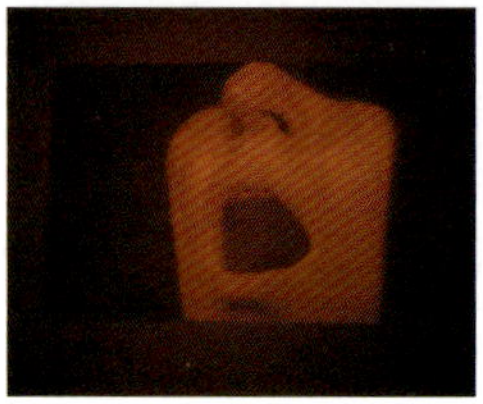
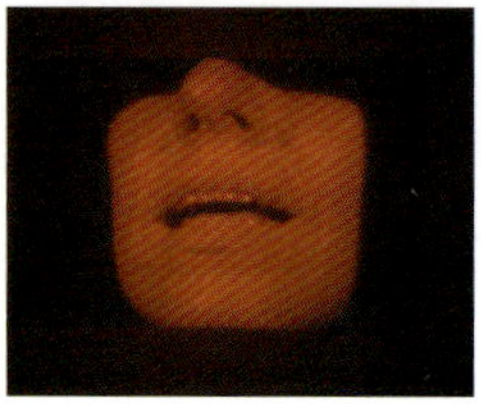
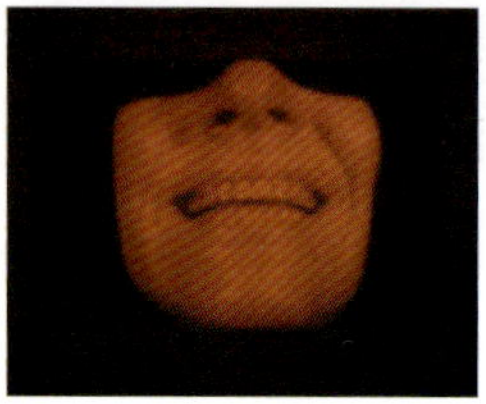
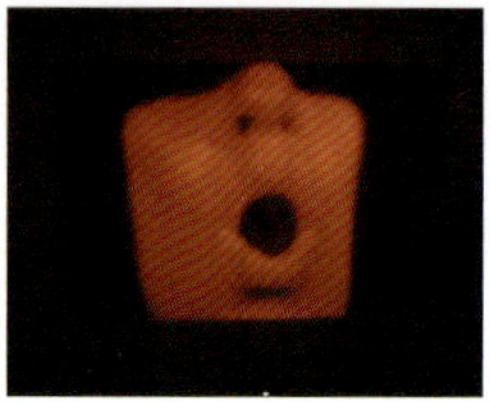
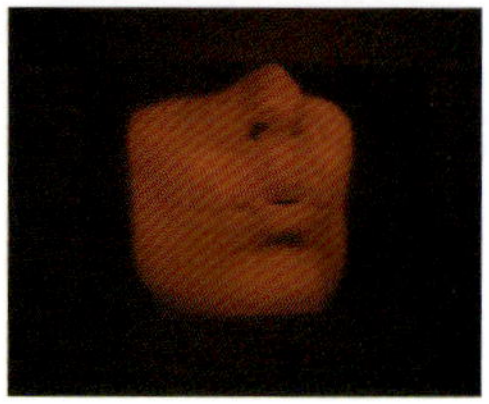

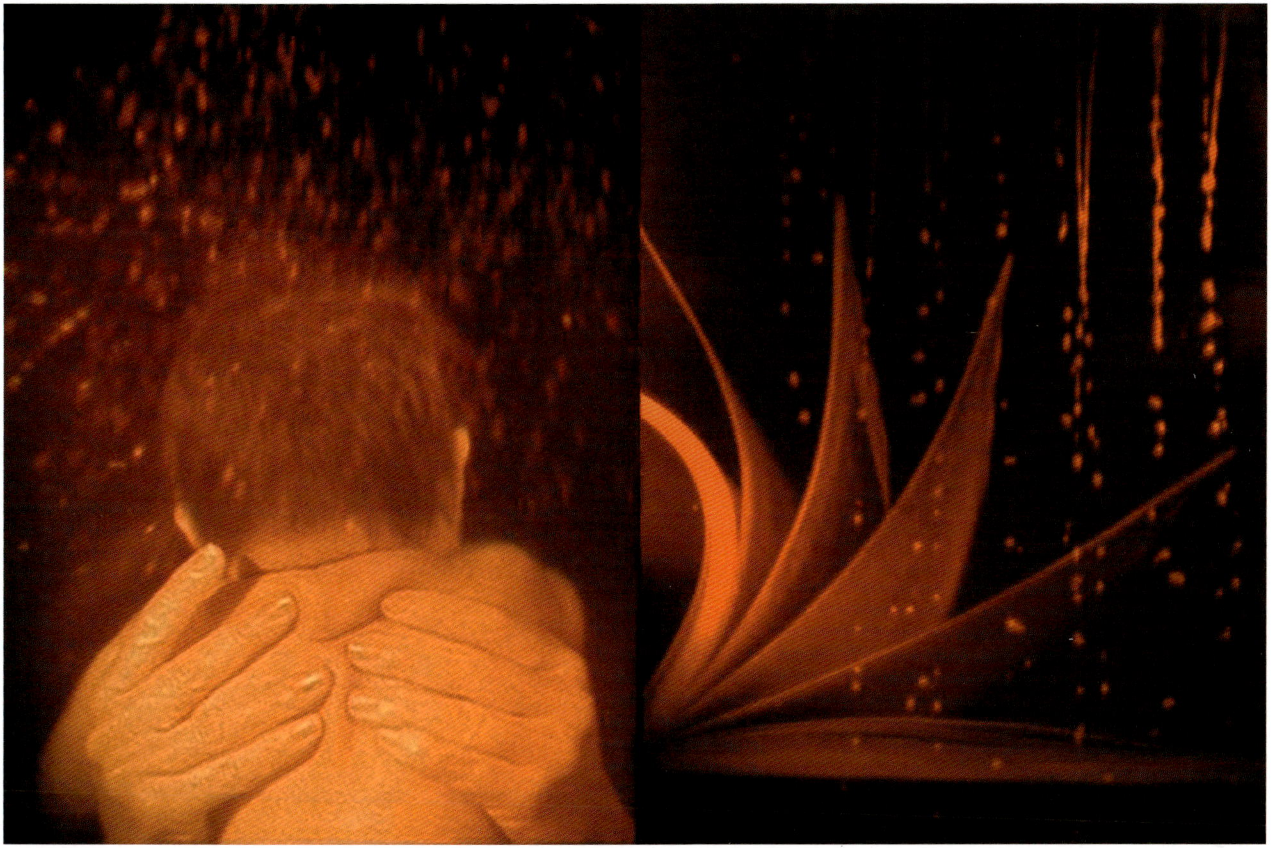

Open book (reading between the lines), above, reflection hologram diptych , 24"x16", 2001

Vowels (breathing), at left, reflection hologram, 16"x12", 2001

The Secret Calculations of Dervish Z

This video collaboration with composer David Weinstein (single channel, 3D glasses, stereo sound, running time: 26 mins.) is a collection of psychological, spiritual and intellectual artifacts presented in three parts. Obsessive numbers and codes emerge from a stream of thoughts as whirling, abstract forms. Mutant forms and audio detritus are fed into a vortex, reduced to digital plasma, and recombined by utilizing ancient techniques known to induce ecstatic states in the viewer. Based on research revealed by the Rand Corporation, the Kabbalah and Nicola Tesla, this project celebrates the promise of spontaneous evolution.

Part 1 Ooze

Awash in a tide of information, the calculations begin with an impulse to order by counting. In turn, the counting exercise works its own effect, leading to the idea of an idea.

Part 2 The Zimmerman Telegram

A secret message which, once intercepted and decoded, thwarted the invasion of the US in 1917.

Part 3 Tzeruf

A kabbalistic practice in which abstract letter combinations were used as keys or entry points to altered states of consciousness. The mystics become aware of meaning beyond meaning flowing into their beings so that their hearts are warmed and they are bathed in bliss as though with rushing water or oil.

When viewers wear the dark lens of the 3D glasses over the right eye, they perceive the depth effect.

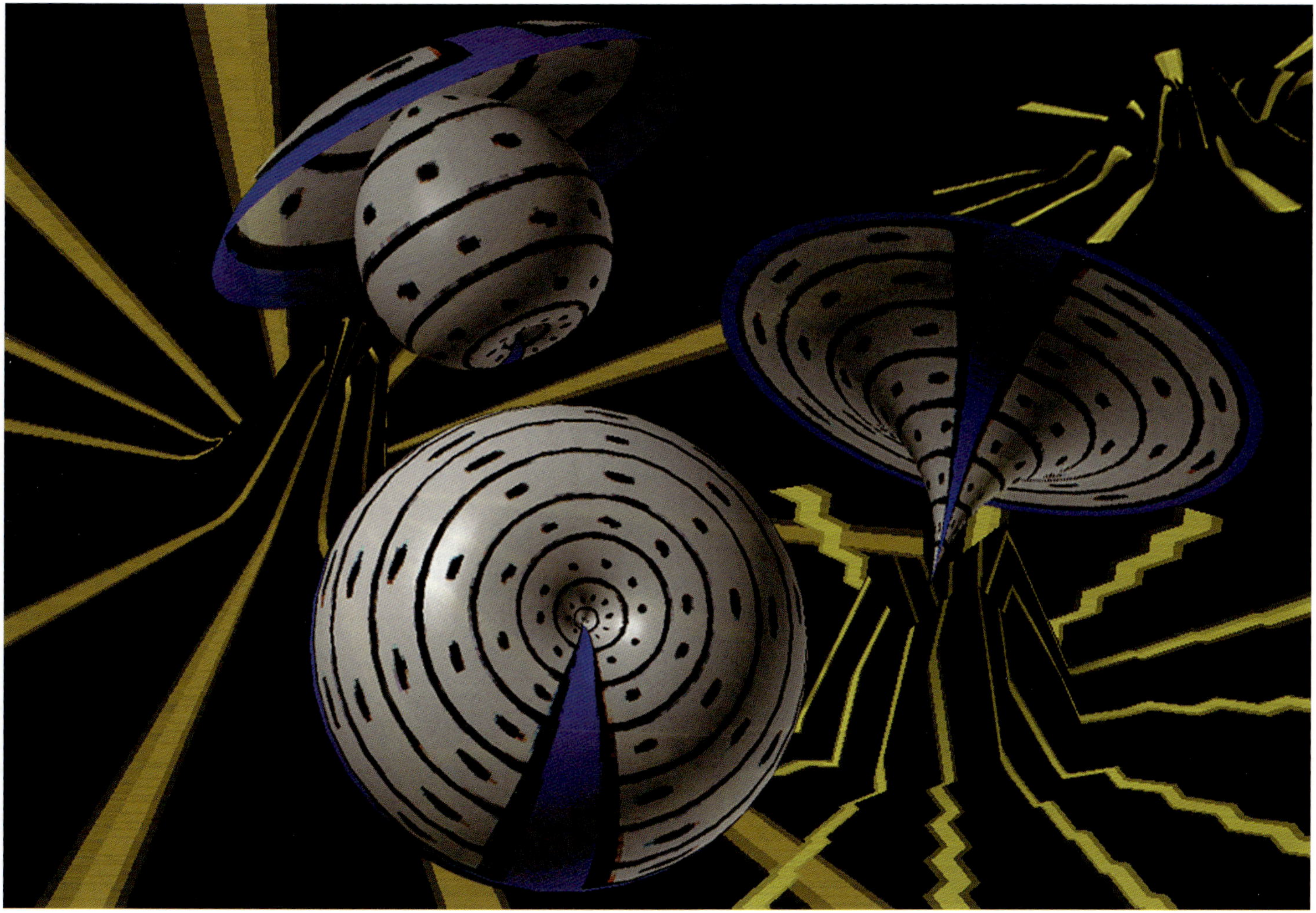

1, 2, 3 much (video, 22 minutes) looks and listens into the fermenting kaleidoscope of artist Adolph Wölfli's work. While jailed in an asylum, he relentlessly calculates the number and weight of his psychic world. He becomes his alter ego, St. Adolph, to compensate for calamities suffered in childhood. His extravagant counting evolves into a numerical chant in a quest for innocence and transcendence.

Born in Switzerland in 1864 to poor parents, Adolf Wölfli was abandoned at an early age. In his 20's, after three attempts to abuse young girls, he was diagnosed as paranoid schizophrenic. At age 30 he was incarcerated in a mental asylum for the rest of his natural life. During the next 35 years, until his death in 1930, he created an astonishing body of creative work, writings, drawings and music, all as he wrestled daily with the voices that haunted him.

Collaboration with David Weinstein, composer and Charles Allcroft, performer, 2003

"...endless eternity is neither round nor square, has absolutely no limits."
– Adolf Wölfli

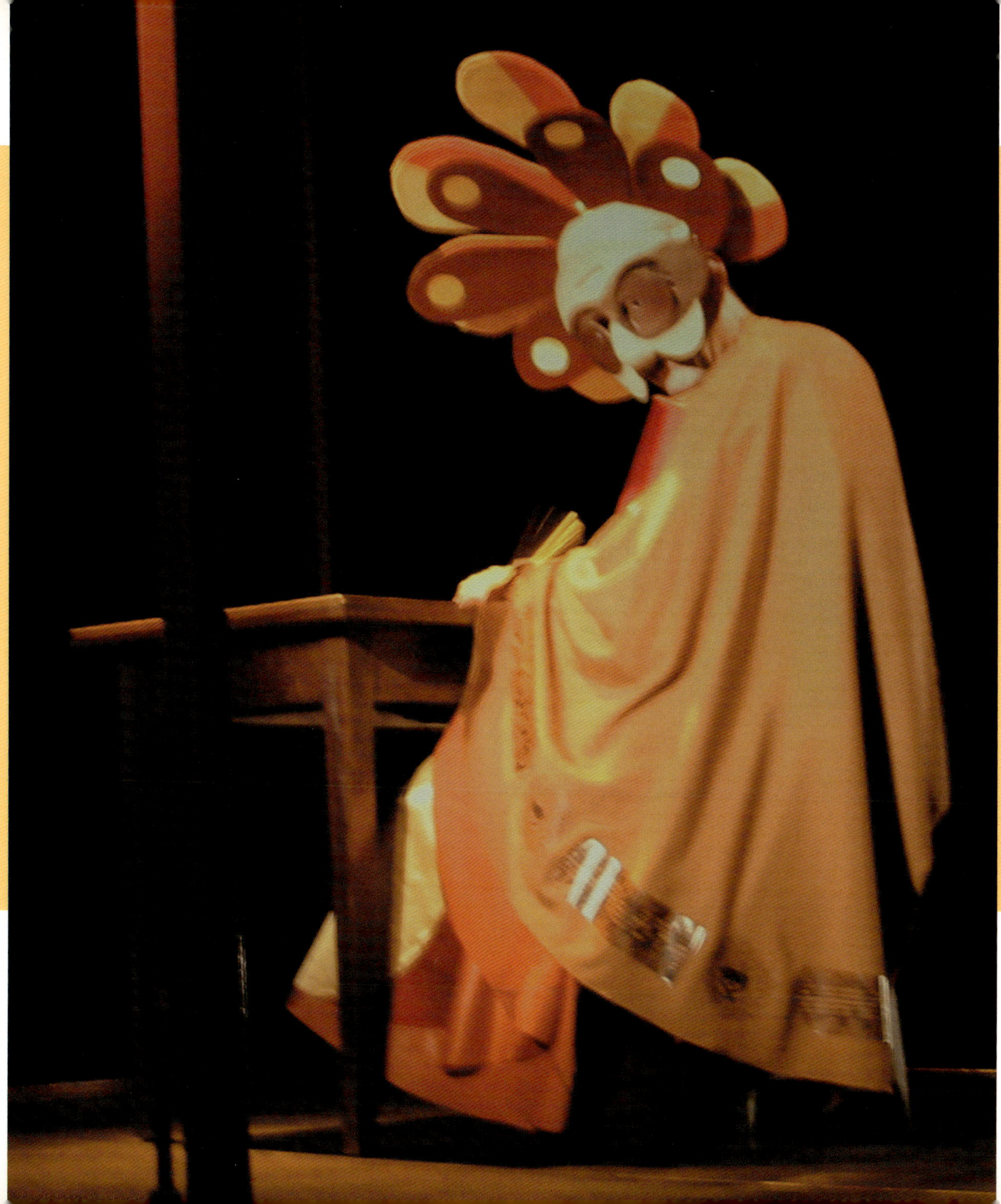

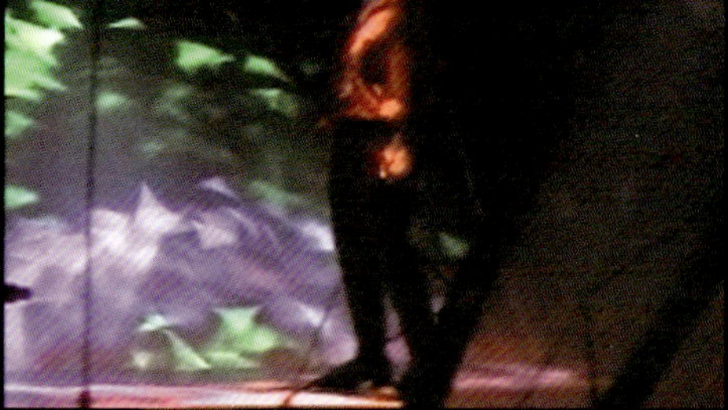
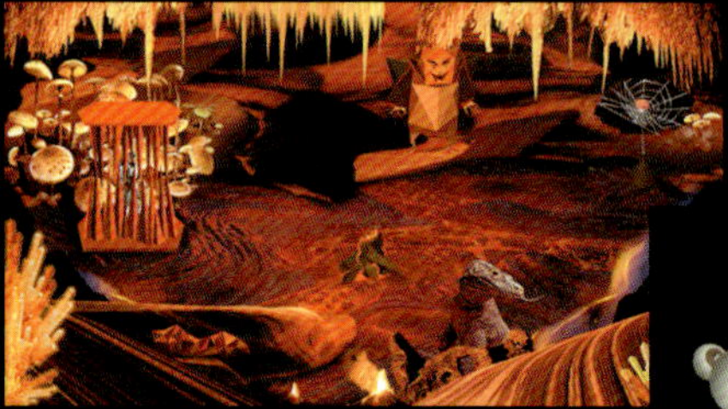
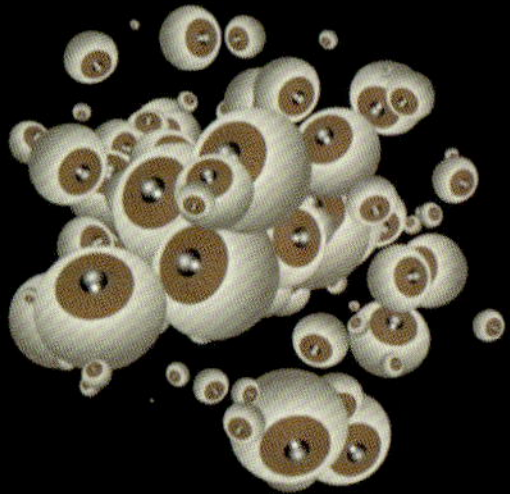
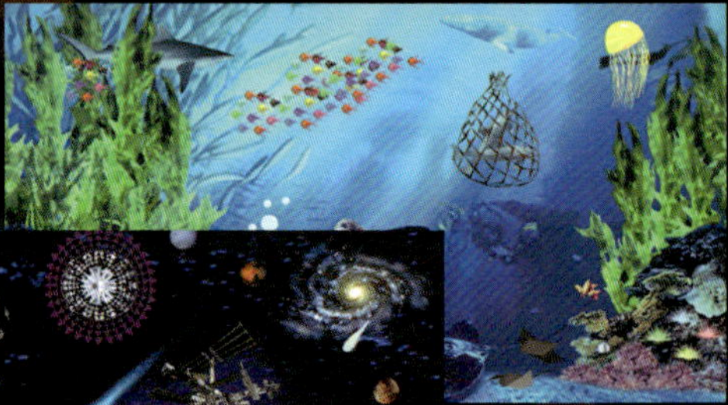
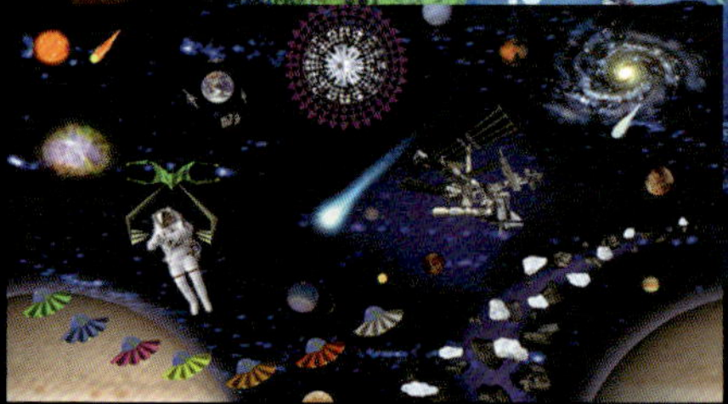

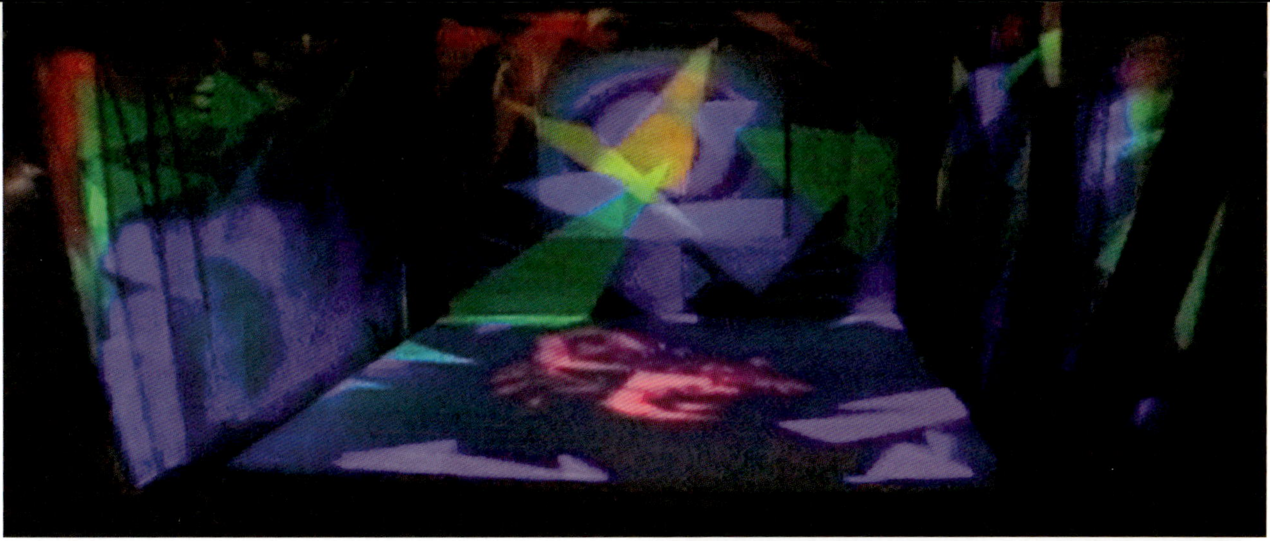

StoryMachine is a learn-play environment where children walk into a big book with ten-foot-high wing-pages. The six-member "crew" picks their way through a series of group-chosen adventures, while the swarming triangles reflect the power-field of their group actions.

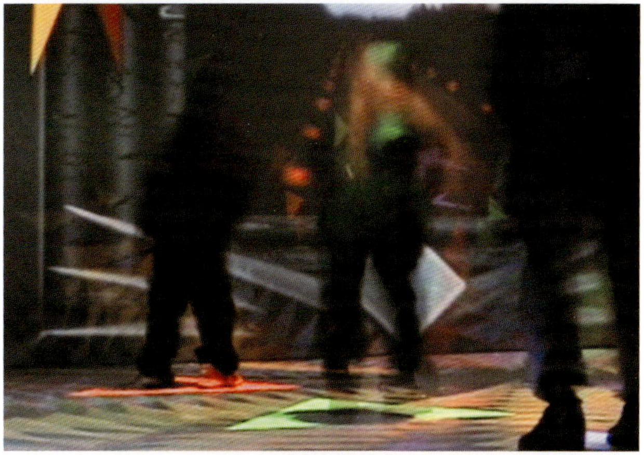

The game invites a group of children to work together, overcoming obstacles to win a goal. The program tracks the children's movement, navigating across large backgrounds. The obstacles and goals are made out of triangles which burst apart and flock around the children when the group calls out together.

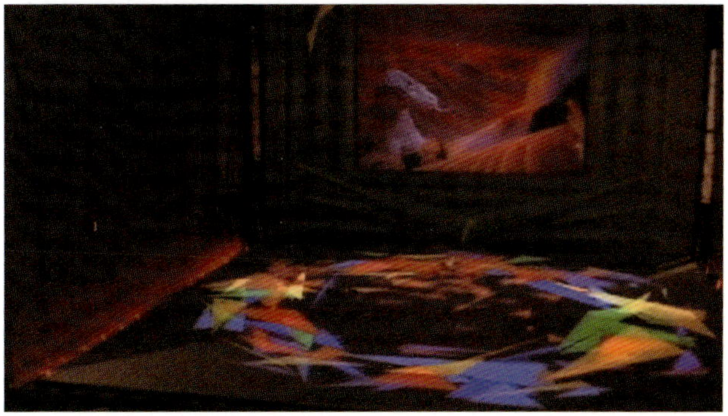

Collaboration with Animax, Bonn, Germany, and David Weinstein, composer, 2000

FX Factory brings kids into the special effects workshop of the teenage inventor, Whoozit. He wants to demonstrate his latest invention but something is out of whack. The workshop floor is filled with animations of machines, which the children explore through group and individual actions. This leads them to a group strategy game to re-connect the machinery. Ultimately, Whoozit presents his invention, the Purple Harmony Engine, which allows the kids to dance with color.

Collaboration with Animax Multimedia Theater, Bonn, Germany and composer David Weinstein, 2001

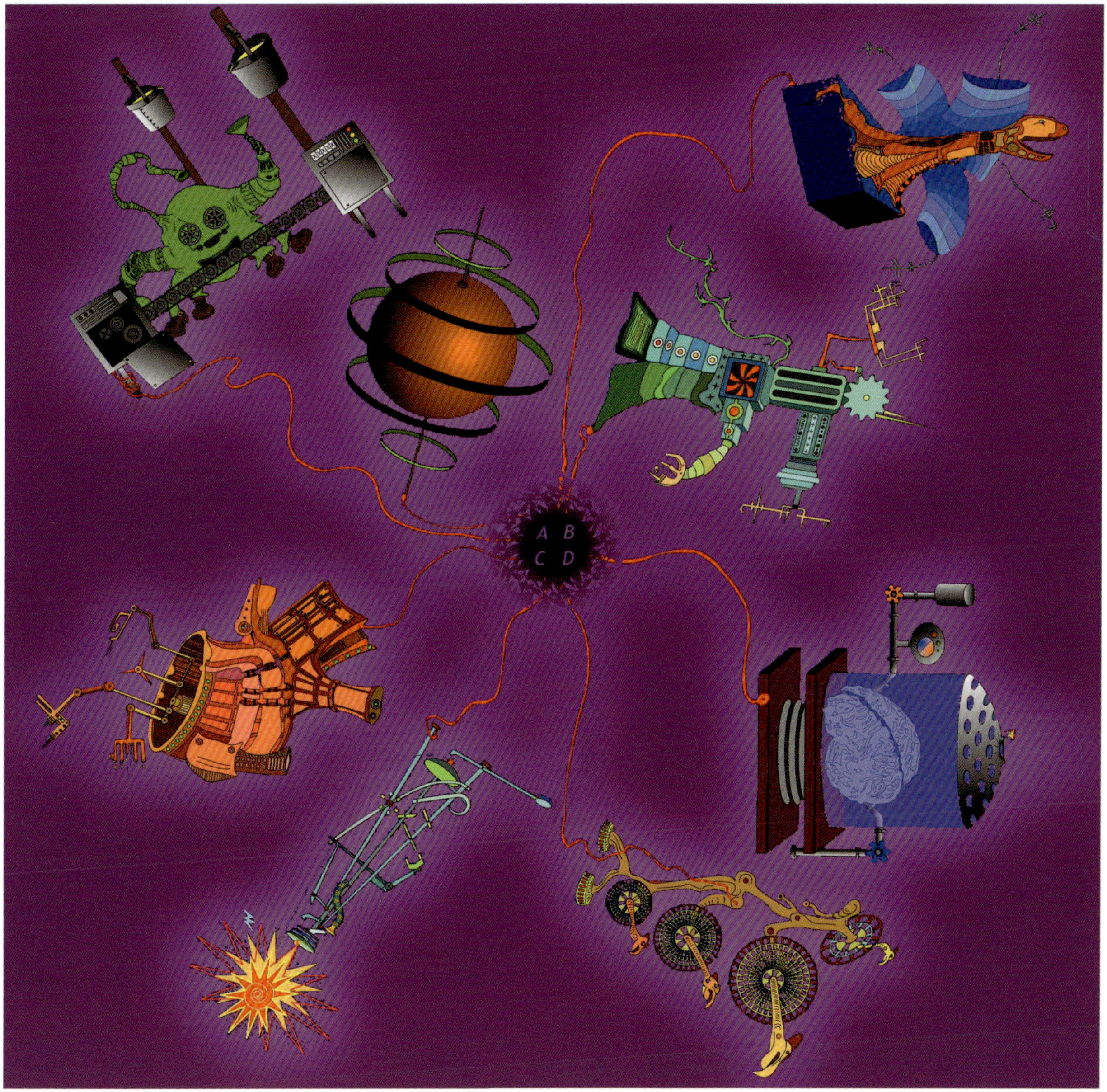
A B
C D

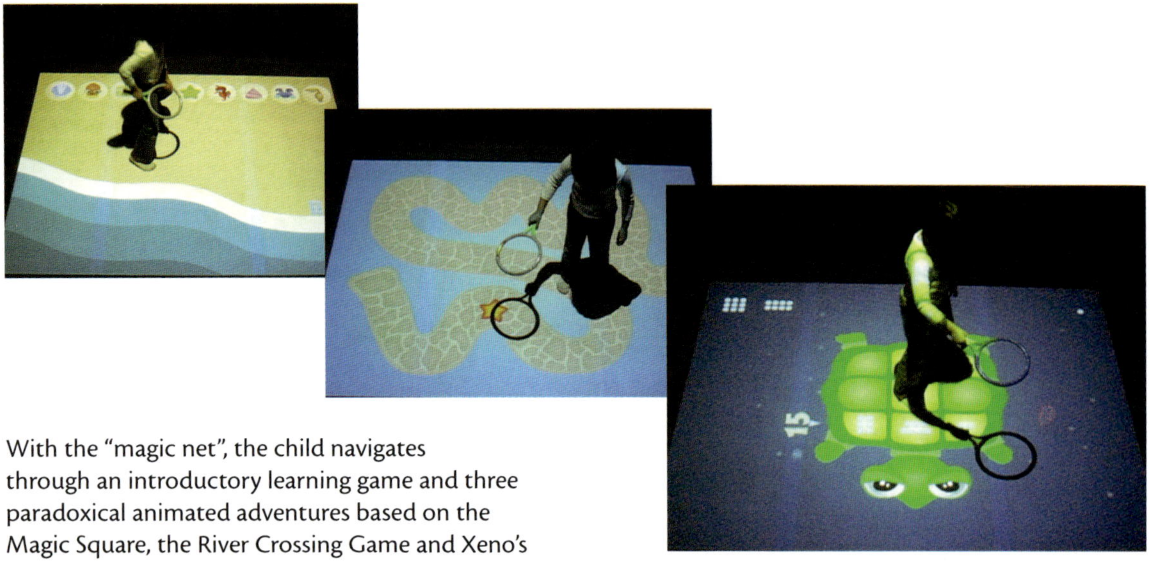

With the “magic net”, the child navigates through an introductory learning game and three paradoxical animated adventures based on the Magic Square, the River Crossing Game and Xeno’s Paradox.

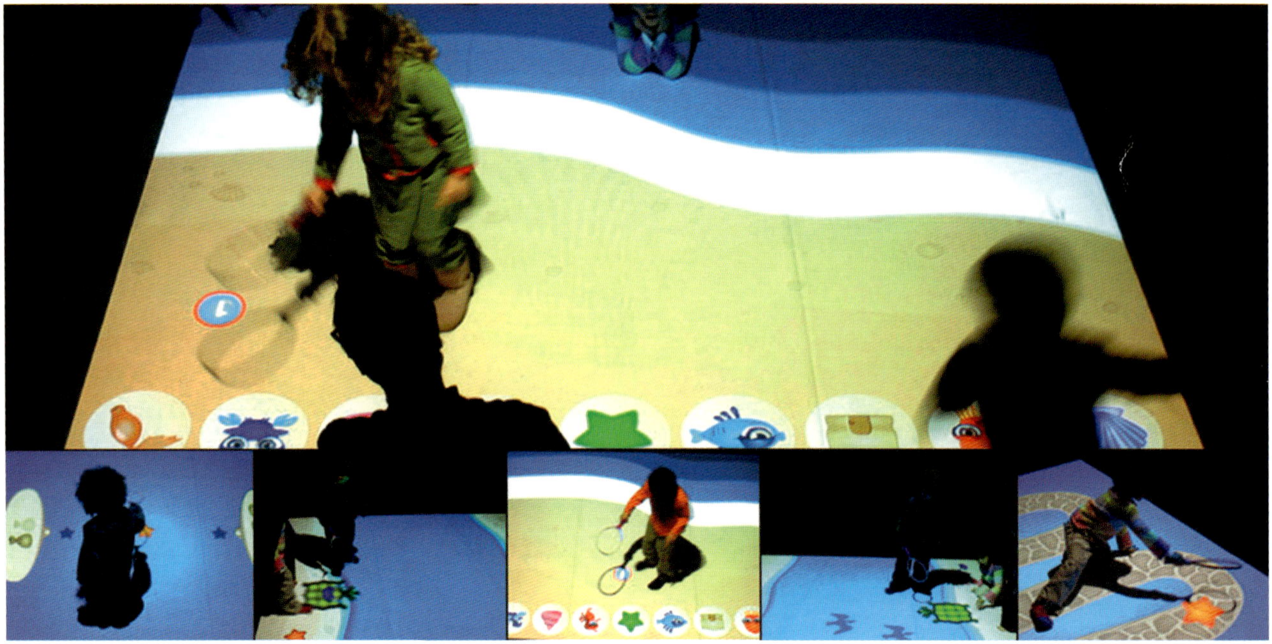

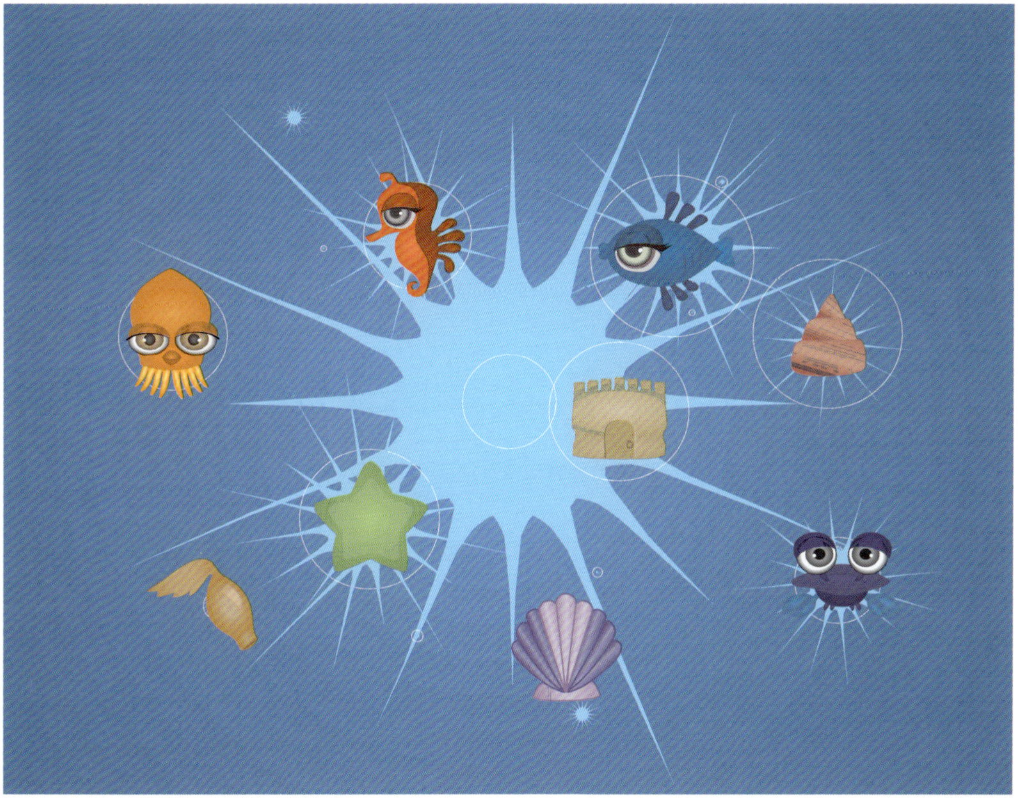

Paradox Beach is a room-sized responsive learning game for kids 6-10 years. Exploring the mystery of numbers, the game begins by washing up numbers like sea treasure on the shore. Children collect the numbers with a magic net and use them to travel through secrets of age-old riddles. A walk-in video playland, Paradox Beach slides into mathematical whimsy, tickling children into solving problems with dance-floor interaction.

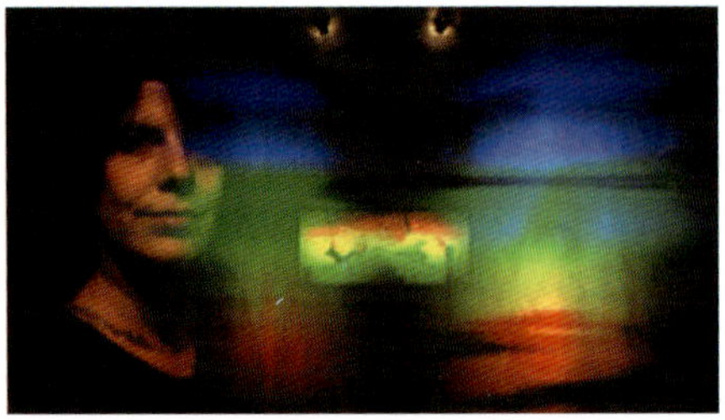

Doris Vila creates luminous environments where stories stretch out in time and space. Often her work revolves around flocks, swarms, and crowds in motion, as viewers trigger sounds, video and holographic scenes.

For over twenty-five years, Vila has worked with art and technology in Europe, Latin America, and the United States. Born in Miami of Cuban heritage, she studied at University of California, Berkeley, and Hunter College in New York City.

Vila writes: "I make holograms and responsive arenas where stories float in space and time. Viewers look and look again, revealing words and objects in a dance of personal discovery. In the weightless story-space, we can walk through the walls between the mind and the senses. In these works the separation between artist, work and audience dissolves.

"One of the key mysteries of a hologram is that it is made of light, weightless light. The image hovers, as if the ghost of the object is caught in a window floating free of the weight of the world."

Her holograms tell of a teasing relationship with science, as she feels that it does not go quite far enough to explore human foibles. Technology and emotion converge in her pseudo-scientific titles. Through found objects, stencils, hand-drawn diagrams and words, her work is light in more than one sense of the word. It's something you believe in because you see it in three dimensions, though it's impossible to touch.

In recent years, with the Vilamedia team, Vila has developed three full-room children's learning games where problem-solving and cooperation are keys to success. Her approach promotes spontaneous group interaction, setting the stage for joint strategies, shared leadership, and, of course, fun.

Visit www.vilamedia.com

SOLO EXHIBITIONS

2008 Every time I open my eyes, I see things, Butler Institute, Ohio
2003 1, 2, 3, much, video on Manhattan Cable
2002 FX Factory, Animax, Bonn, Germany
2000 StoryMachine, Animax, Bonn, Germany
1999 Wölfli, A journey into chaos, Cologne, Germany
1995 Theatermachine, Tapetenfabrik, Bonn, Germany
A Flock of Words, NYU, New York
1993 The Book of Air, Chip Arts Atelier, Bonn, Germany
(A self-help survival guide to the) Global Village..., Germany
1989 Space, time, sex, money & holographic stories, Museum für Holographie & neue visuelle Medien, Germany
Heaven, home and weightless, Photographic Resource Center, Boston, Massachusetts
1988 Holographic Narratives, Interference Gallery, Toronto
1984 Holografías de Doris Vila, Facultad de Artes Plásticas, Universidad Veracruzana, Mexico

SELECTED GROUP EXHIBITIONS

1999 Lumia, Charlottenborg, Copenhagen, Denmark
1998 Transmediale, Berlin, Germany
1996 15 Lux Maximum, New York, NY
Akademie der Künste, Berlin, Germany
1995 Art & Technology, State University of New York, Stonybrook
1993 Holographic Exhibition, Nagoya, Japan
Light Dreams, Atlanta Museum of Science & Technology,
Kultur und Wirtschaft, Glucksburg, Germany
1992 The Edge of Light, Museum of Holography, New York
1991 Räume aus Licht, Akademie Galerie, Berlin, Germany
Eye Believe, Science Center, Des Moines, Iowa
1990 Vistas Latinas, New York, NY
Light Dreams, Canton Art Institute, Ohio, and Pensacola Museum of Art, Florida
Captured Light, Museum of Holography, New York
1989 World of Holography, Chunichi Shimbun, Nagoya, Japan
Visiona, Präsidialabteilung, Zurich, Switzerland
As We See It, National Geographic Society, Washington, DC
1988 Images du Futur, Expotec, Montreal, Canada
BIG Holograms, Los Angeles, California
American Art Holography, Museum of Holography, New York
International Display Holography, Lake Forest, Illinois
1987 Images in Time and Space, Montreal, Canada
Interface 87, Art/Technology, Morris Museum, New Jersey
Holography Show, Nexus Foundation, Philadelphia, PA
1986 Windows on White, installations, New York City
Four New York Holographers & Innovation through Tradition, Museum of Holography, New York
1985 Holography, School of the Art Institute of Chicago, Illinois
International Display Holography, Lake Forest, Illinois
CAPSIZED, BACA, Brooklyn, NY
Perceptions, Interference Gallery, Toronto, Canada
1984 Lichtblicke, Deutsches Filmmuseum, Frankfurt, Germany
Holography (Re)Defined, Museum of Holography, New York
New Light, Chicago Cultural Center, Chicago, Illinois
1983 Falling, BACA, Brooklyn, New York
Seeing Double, Stereo Imagery, Visual Studies Workshop, Rochester, New York
International Film Expo, New York
1982 Mimes and Minatures, BACA, Brooklyn, New York
1979 Muestra Colectiva, Parque Juárez, Xalapa, Veracruz, Mexico
Plástica de Xalapa, Institute of Fine Arts, Mexico, DF
1978 Concurso de Aguascalientes, premiada en pintura, Palacio Nacional de Bellas Artes, Mexico City
1977 Tecnología Alternativa, Galería Orozco, Mexico, DF

COMMISSIONS AND COLLECTIONS

Agfa-Gevaert, Inc.
Museum of Holography, New York.
Chunichi Shimbun, Japan
Museum für Holographie und neue visuelle Medien
GMD, Germany
New York University Medical Center
Interferens Holografi, Norway
Sammlung Fielmann
MIT Museum, Cambridge, Massachusetts
University of Wisconsin
Museum of Holography, Fine Arts Research, Chicago
Washington State Arts Commission

AWARDS

Ars Electronica, 2002, Special Jury Mention, Interactive Art
Center for Holographic Art, 2000
Life 2.0, international competition for artificial life art, 1999
Artist in Residence, GMD, German computer research center, 1993
InterActiva Symposium Award, for interactive holographic video environment work, 1993.
Fellow, Academy of Media Art, Cologne, Germany.
Fellowship, MacDowell Colony for the Arts, 1986, 1988, 1991, 1998
Shearwater Foundation Grant, 1988
Cintas Fellowship, 1986
Inaugural Artist in Residence, Global Images Laboratory, 1985
CAPS Grant, New York State Council on the Arts, 1984
Artist in Residence, Museum of Holography, 1983, 1988
Visiting Artist, MIT Regional Laser Center, 1982

SELECTED WRITINGS

Chasing Rainbows: One Holographer's Approach, LEONARDO, VOL. 22, NO. 34, PP. 345-348, 1989
Among many technical papers:
Folding Time into Space, Proceedings of SPIE, International Society for Optical Engineering, Vol. 1914, 1993.
Holo-dynamics, Proceedings of SPIE, International Society for Optical Engineering, Vol. 2716-20, 1994

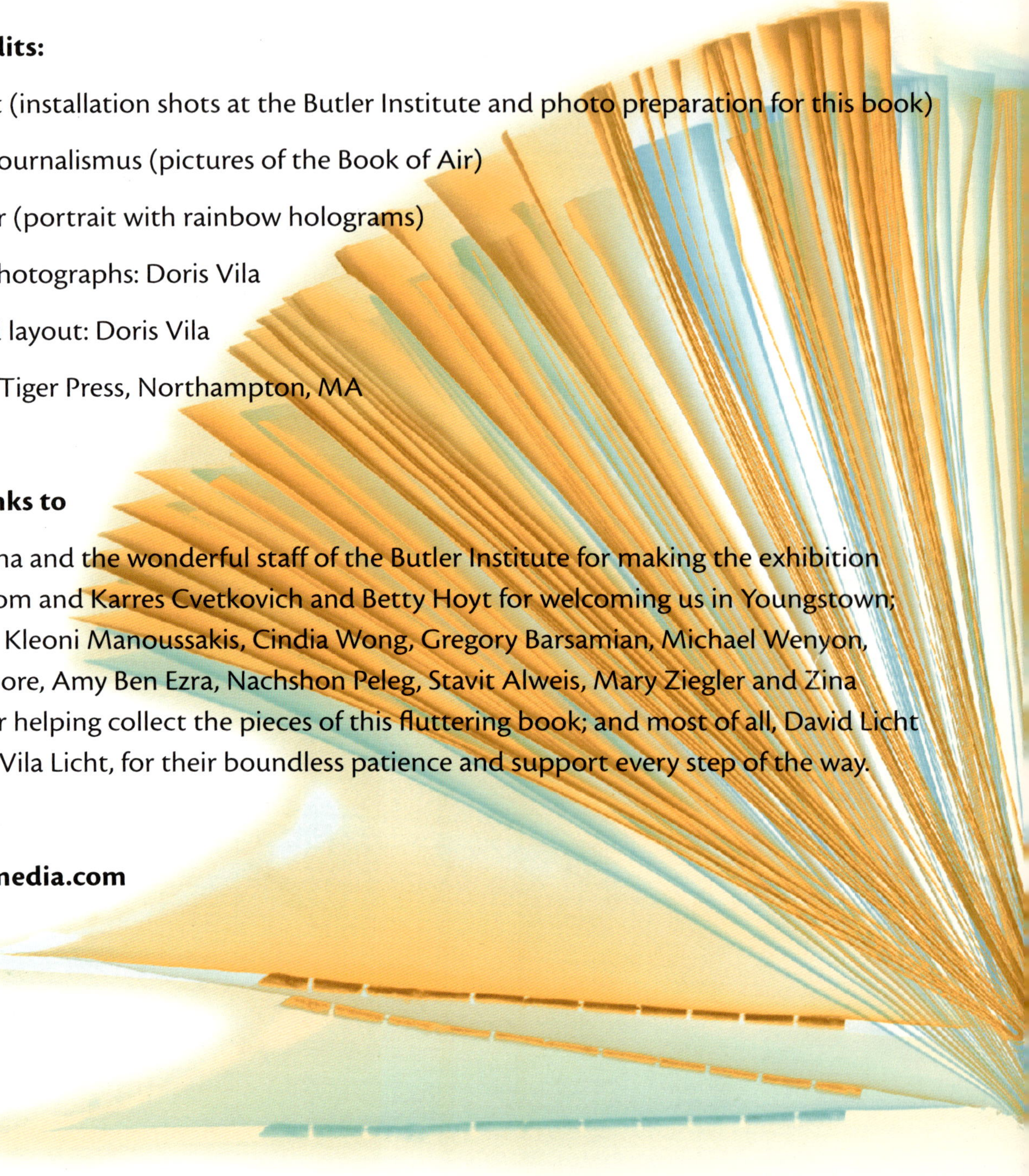

Photo credits:

David Licht (installation shots at the Butler Institute and photo preparation for this book)

Joker Fotojournalismus (pictures of the Book of Air)

Sally Weber (portrait with rainbow holograms)

All other photographs: Doris Vila

Design and layout: Doris Vila

Printed by Tiger Press, Northampton, MA

Many thanks to

Dr. Lou Zona and the wonderful staff of the Butler Institute for making the exhibition possible; Tom and Karres Cvetkovich and Betty Hoyt for welcoming us in Youngstown; Ken Butler, Kleoni Manoussakis, Cindia Wong, Gregory Barsamian, Michael Wenyon, Sandra Moore, Amy Ben Ezra, Nachshon Peleg, Stavit Alweis, Mary Ziegler and Zina Goodall for helping collect the pieces of this fluttering book; and most of all, David Licht and Lillian Vila Licht, for their boundless patience and support every step of the way.

www.vilamedia.com